I0803120

IMAGES
of America

NORTHPORT

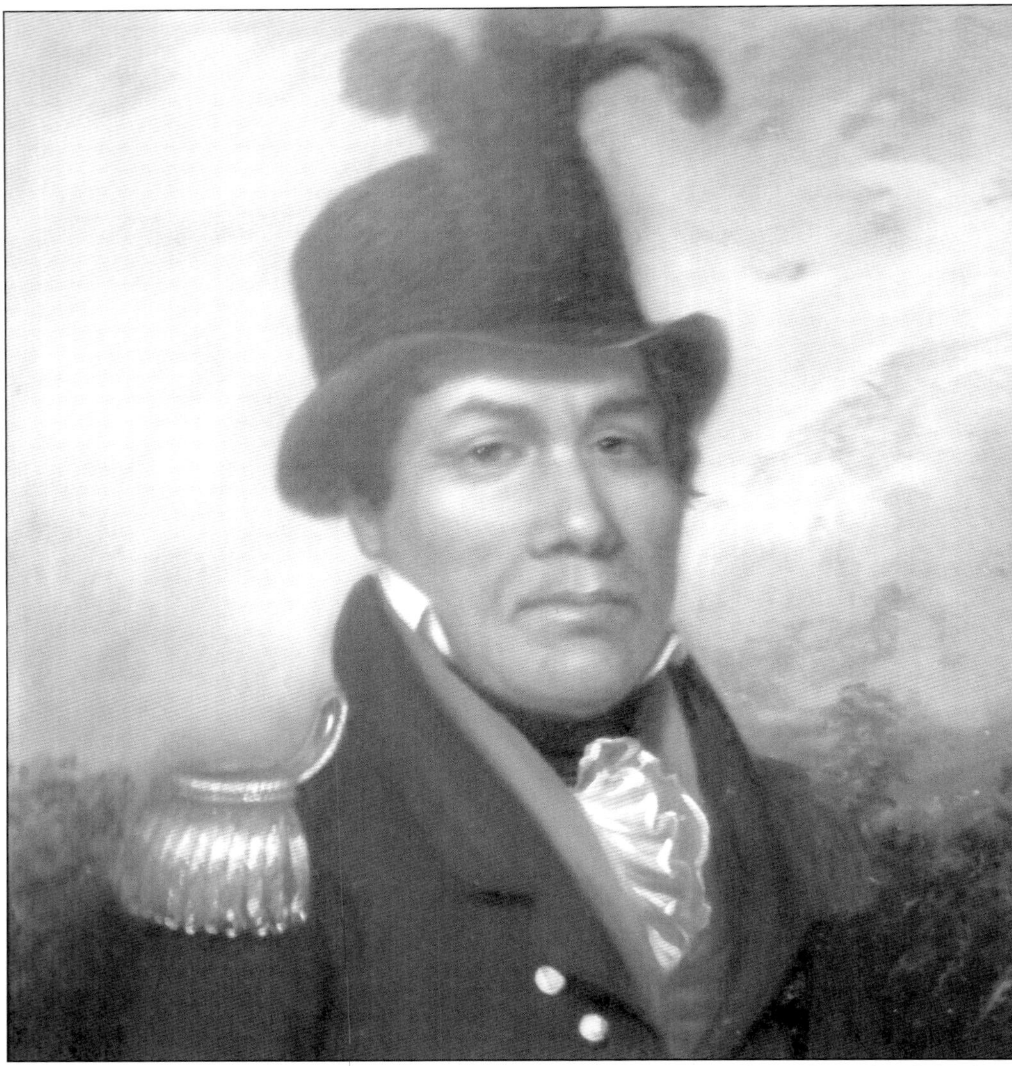

In the early 1780s, the Northport riverfront west of the train trestle and south of the levee was the scene of a fierce battle between the Muskogee (Creek) and Chahta (Choctaw) Indian tribes. During this battle, the legendary Pushmataha was instrumental in the Choctaw victory. It was on this site in Northport that Pushmataha earned his warrior's name. According to the account written by early Northport settler Gideon Lincecum, 27 Choctaws were buried on this site along with at least 20 Muskogees. In addition to the historic battle, this was also the site of the original settlement of Northport, known in the early years as Canetuck, or Kentucky. Marr's Ferry Road, built about 1820, still exists on the property. (Courtesy of The Westervelt Company. *Pushmataha*, by Charles Bird King, housed in the Tuscaloosa Museum of Art; home of the Westervelt Collection.)

On the Cover: Orange Crush Bottling Company was founded in the early 1920s by William B. McClendon. The distribution building was on the south side of Tenth Street across from Northport Baptist Church. Among the products produced by McClendon was Try-Me. An advertisement for the beverage stated, "Try-Me beverages are not habit forming, but you'll get the habit because they're good." (Courtesy of Jane McClendon Ball.)

IMAGES
of America

NORTHPORT

Friends of Historic Northport, Inc.

ARCADIA
PUBLISHING

ISBN 978-0-7385-9400-2

Published by Arcadia Publishing
Charleston, South Carolina

Printed in the United States of America

Library of Congress Control Number: 2012942637

For all general information, please contact Arcadia Publishing:
Telephone 843-853-2070
Fax 843-853-0044
E-mail sales@arcadiapublishing.com
For customer service and orders:
Toll-Free 1-888-313-2665

Visit us on the Internet at www.arcadiapublishing.com

This book is dedicated to Marvin L. Harper, local preservationist and founder of Friends of Historic Northport, Inc.

Contents

Acknowledgments

In 2000, Marvin Harper first introduced me to the history and heritage of Northport. Through his encouragement, I became a member of Friends of Historic Northport, Inc. (FHN) and have served on its board of directors for the last 12 years.

Thank you to Carl Adams, Lackey Stephens, and Hayse Boyd. Their research assistance has been invaluable in making this book as accurate as possible. Without their encouragement, ideas, and wealth of information, this effort might never have been completed. Amy Materson provided editorial assistance and kept me motivated to complete the book. Betty Booth, June Lambert, and Frances Rice Pool proofread the text for grammatical and historical errors.

Over the years, many FHN members have taken and donated photographs and written photograph captions. Among those members is Lige Moore, the official photographer for FHN for many years. Sara Brunette managed the photograph collection, and later, Patricia Thrasher took on those duties. Thank you all for your hard work.

Another source of information was FHN's partnership with the Heritage Commission of Tuscaloosa County. Use of their collections in researching the history of Northport and the surrounding area has been invaluable.

The best photographs were chosen from the resources available. While a particular person or place may not be shown due to space limitations, those individuals, structures, and memories are not forgotten, and are carefully preserved.

Unless otherwise noted, images are courtesy of the Friends of Historic Northport, Inc. Photographic Collection.

Thank you to Arcadia Publishing, including Simone Monet-Williams and Maggie Bullwinkel, for helping me through the process of putting materials together and for being advocates for me in support of our collective effort to publish this book.

—Chuck Gerdau

INTRODUCTION

Northport is located on the northern side of the Black Warrior River across from Tuscaloosa. Before Alabama became a state, it was the site of a convenient ford for travelers crossing through the Alabama Territory. The river was navigable for a portion of the year up to Tuskaloosa Falls, where Northport stands today. Because it was the northernmost point on the Black Warrior that could be used for commercial transportation, it was a natural site for the region's main port.

James Christian and Capt. Otis Dyer were among the city's founders. Captain Dyer operated a ferry between Northport and Tuscaloosa and was instrumental in laying out Northport's streets and lots. Much of western Alabama's cotton crop passed through the city's port on its way to New England and beyond.

Northport was first known as Canetuck, or Kentucky, because it was surrounded by a dense canebrake wilderness. It later became known as North Tuskaloosa. By the early 1840s, the name North Port was used by the post office. Throughout the 19th century, all of the courthouse records used this spelling.

In the early years, North Port and Tuskaloosa had similar facilities for handling the cotton brought to the river from the vast territory north and west of the towns and for shipping and receiving freight by the river.

Northport's earliest businesses were along the banks of the Black Warrior River. Early settler Col. E.A. Powell wrote that, "The town consists of a squatty looking building right on top of the river bank . . . with a long piazza across the front . . . It was a gambling hall and a place of great disturbance . . . Three buildings represented the commerce on the north side of the river, and the western end of the little town was a canebrake."

In the mid-1830s, due to the flooding of the Black Warrior River, Northport businesses began to move north away from the river. The new town center grew around the intersection of Columbus Street (now Fifth Street) and Main Avenue. In addition to cotton warehouses along the river, including those of Elias Persinger, who arrived around 1830, several frame store buildings were built on the west side of Main Street, from Columbus Street southward.

Around 1850, a great fire burned most of the downtown, much of which was replaced with eight or ten fireproof brick buildings built by James Shirley. One of those buildings, the Northport 5 & 10, still stands today.

During the Civil War, Northport was damaged extensively by Croxton's Raiders in April 1865. Upon leaving Tuscaloosa, they burned the bridge connecting it to Northport. During Reconstruction, many citizens migrated to places such as Texas, and Northport struggled for many years to reestablish its economy.

Northport was incorporated in February 1871. The boundaries of the town went from the boundary marker along the Black Warrior River west to Dempsey Williamson's property, north from there to a turn at the Academy Hill and on to the old Bridge Avenue, where it turned south again to the river.

An 1889 Sanborn fire map of Northport shows businesses lining both sides of Main Avenue between what was then First and Columbus Streets, many of them listed as "general merchandise" stores. There were also two or three grocers, a vegetable market, a meat market, a drugstore, a jeweler, a hardware store, and the large McGee & Son cotton warehouse. Behind many of the stores were warehouses and horse and mule sheds.

The current buildings on Main Avenue were primarily built after the Civil War, with the stores on the east side rebuilt after an 1891 fire. The first telephones were installed in 1894, and three years later, electric lights became available for home use. In 1898, the Mobile & Ohio Railroad began operations in the city.

Despite its unpredictability, the river aided the continuing development of Northport. The first lock and dam was completed in 1896, which, along with subsequent lock and dams built in the following years, provided a major source of employment.

Aiding the economic boom was the construction of the Mobile & Ohio railroad bridge across the river. The railroad allowed businesses to send and receive goods that were not available via the river. Along with the new Northport river dock, the railroad helped the city become a major shipping port for goods heading south to Mobile and onward. Cotton farmers, merchants, and others brought goods from all over Tuscaloosa County and from Fayette, Pickens, Lamar, and Walker Counties to be shipped.

The abundant fish in the Black Warrior River created a booming fishing industry, which fed and provided jobs for residents. An 1896 *Tuscaloosa News* article stated, "For years past it has been a common thing to see wagon loads of fish in our streets, and the over-plus stock is now shipped to neighboring cities and towns . . . Fish are caught in various ways, by lattice traps on the reefs, fish baskets, nets, trot lines, hook and line, etc."

In 1900, Northport had a population of between 600 and 800, and while the waterway provided food and cheap transportation, it also flooded the town and the surrounding region with surprising regularity. At the time, wagons and buggies filled the often-dusty streets and horses were tied to posts outside businesses and residences, but things changed in February 1904 when Tuscaloosa County probate judge James C. Brown issued the first county automobile license to Northport merchant W.S. Persinger. As late as 1912, automobiles were still unusual enough that new ones warranted mention in the newspaper; one article reported, "Mr. S.P. Faucett has purchased an auto. This makes eight cars for Northport." In 1916, a 10-mile-per-hour speed limit "around curves" was established for "fast autoists."

One particularly prosperous business was the Alabama Slat and Lumber Company. It was established on Compress Street (now Thirtieth Avenue) in the 1890s, and in 1906, Arthur Laycock reorganized and greatly expanded it.

Coal-mining operations were also major employers in the area. Most of the local coal, discovered in the mid-1800s, lay close to the surface, making the majority of excavations strip mines. However, one deep-well-shaft mine was near what is now Tuscaloosa's Indian Hills subdivision and a few others were northeast of town. Today's Rice Mine Road is named for a road that led from Northport to the Rice brothers mining operation in the early 1900s.

There were several coal mines farther away from town, including Standard Coal Mining Company, near Brookwood. The coal and income produced by the mines helped the city grow and flourish. Almost all the homes and businesses in the city were heated with coal, with many homes having three or four heaters or fireplaces fueled by local coal. Many of the area's cook stoves also ran on coal.

In 1900, the town, with a reported annual revenue of $1,500, had only one full-time employee, a town marshal who was paid $30 a month. That year, the mayor's salary was $75 a year, and councilmen earned $25 per year.

In the early 1900s, Northport had a two-room school on Ninth Street near Main Avenue and electricity was beginning to light homes and streets, replacing wood and coal as a fuel source.

In 1908, the city established a major ordinance implementing fees and city taxes to produce income. At that time, obtaining drinking water was still largely a matter of personal initiative for the city's approximately 800 residents, with private wells supplying most homes. A main well and scales were located at the intersection of Columbus Street and Main Avenue. By 1915, there was a limited city water system in use, supplied by a spring located on Compress Street (now Thirtieth Avenue).

In 1914, a fire leveled parts of downtown, primarily the west side of Main Avenue. The block was built to be fireproof after the 1850s fire destroyed the original log and frame structures, but unfortunately, the fire prevention alleys were filled in by frame structures, allowing the street to burn once again.

In 1920, Northport's population was listed at 1,606. In the middle of that decade, Northport city government moved into the upper level of the two-story Shirley Building on Main Avenue. A barbershop was on the ground floor and there were exterior stairs on the side of the building leading to the city offices, which remained there for almost 30 years.

The 1930 census listed the population at 2,173, but as the Great Depression gripped the nation during the following decade, Northport suffered economically and physically. However, in 1934, the Works Progress Administration completed a new Northport city jail. By the end of the 1930s, the prewar economic boom loosened the stranglehold of the decade-old Depression.

By 1940, the city's population had grown to 3,187, and a few years later, World War II brought an increase in traffic to the Black Warrior River. Barges laden with raw materials for the war effort regularly passed through the river's locks. The railroad also reflected the mass movement of men and machines from one coast to another. Trains loaded with tanks, artillery pieces, and jeeps were as normal a sight during those years as the ever-present "troop trains."

An August 24, 1949, a *Tuscaloosa News* article entitled "Story of Northport Shows the Town's Growth Since 1900" offered a glimpse of the town from its beginnings until mid-century. A photograph titled "Northport Main Avenue in 1885" showed a dirt street, horses, buggies, and wagons, and some of the street's business buildings. The photograph's caption read, "Largest building on the right is the original frame structure for George W. Christian's store, the site now being occupied by Christian & Son. Notice on the left the public well and the public scales. Citizens shown in the picture include Lucien Strong, Sam Faucett and John Anders."

Another photograph showed "Northport Main Avenue Today." According to the caption, the photograph gave "just a hint of the size and scope of Northport's business section." The caption also told of the city's tremendous growth and prosperity: "Progressive steps have been taken in rapid succession within the last decade. Street paving has continued apace. Northport now had approximately 15 miles of paved streets, many blocks being completed this summer. Sewers were also laid for many blocks."

The remainder of the article focused on Northport's civic and political leaders and their accomplishments, including the 1942 purchase and subsequent development of Northport City Park. The article finished with a rosy picture of the city's future: "Upon the firm triangle of the church, the school, and the enlightened citizenry, Northport has marched far since she chose the upward path of progress for a community of opportunity and achievement."

The 1950s were good years for Northport, which started the decade with a population of 3,885. In 1954, the city hall moved to the building occupied by Palmer Realty Company on Fifth Street. Separate quarters were set up in the rear for the police department and a two-way radio system was installed for the first time. Two patrol cars were put into action—up from just one—and a desk officer went on duty 24 hours a day.

In 1956, Northport won a second-place award in the National Community Achievement Contest. That same year saw the completion of Crestmont Elementary School and the Northport branch of the First National Bank, the dedication of the Powell-Shamblin National Guard Armory, and the erection of 120 street markers in Northport.

In 1959, building soared as Piggly Wiggly opened in one of the largest business buildings in Northport. Marion Smith also built a 14-unit motel and Dr. Norman Carlson built Northport's first dental clinic. In the Northwood Lake subdivision, 46 new homes were built. New residential subdivisions were also opened by John Smelley and D.T. Plott on upper Main Avenue, and several homes were built in a subdivision east of Bridge Avenue opened by E.J. Shipp.

In 1960, when the population stood at 5,245, the first structure used exclusively as a post office was dedicated in November on East Fifth Street. Later in the decade, the city annexed several acres, adding more than 1,900 people to Northport—accounting for nearly half of the city's population growth during the decade.

A new city hall was built in 1964, and in 1965, Northport established its first fire department. In 1968, a branch of the Friedman Library opened downtown, and that same year, Northporters used voting machines for the first time. The city's population nearly doubled in 10 years, and in 1970, there were 9,435 people in Northport.

The 1970s saw especially vigorous expansion, with some of the city's growth attributed to the attention brought by Northport's 1971 centennial celebrations, which involved the entire community. The Kentuck Festival of the Arts also grew out of the 1971 celebration. The annual affair has enjoyed tremendous growth, and its widespread reputation for excellence continues to flourish. The event is recognized for its eclectic array of art forms and variety of musical talents.

The Hugh Thomas Bridge over the Black Warrior River was completed and opened for traffic in 1973. The hospital, now called Northport Medical Center—the city's largest current employer—opened in June 1976.

By 1980, Northport's population had grown to 14,291, and by 1982, city hall was so overcrowded that city leaders decided to relocate it from the corner of Lurleen Wallace Boulevard and Twentieth Street. In November of that year, Northport mayor Frank Manderson and the city council enacted a 1¢ city tax to help build a new city hall. A site was chosen and construction began in April 1984. On April 21, 1985, the city formally dedicated the Northport Civic Facility, on Highway 82. The 41,600-square-foot building cost nearly $1.4 million.

The 1980s saw the population grow by more than 3,000, to 17,366 in 1990. In 1992, Northport Medical Center became part of the DCH Medical Group. Also in the 1990s, several buildings of historical value were acquired by the city or by one or more of its active preservation groups: the Friends of Historic Northport, the Northport Renaissance Commission, and the Northport Heritage Commission. The 1838 Shirley Place is now a museum and gardens and the 1907 Palmer House serves as the Northport Heritage Museum.

Community organizations, hand in hand with local businesses and merchants, focused on highlighting the history and heritage of Northport in their efforts to improve the area. Strong interest and commitment from local civic and governmental groups have kept the community moving forward.

One

Welcome to Northport

Welcome to Northport. Photographs that best symbolize and describe our history and heritage have been chosen for this chapter. Descriptions of each image have been written so that a person, new to the community, can quickly envision times gone by while standing in the center of town—Fifth Street and Main Avenue.

In 2012, Dr. Hayse Boyd, a lifelong Northport citizen and accomplished history researcher, completed a detailed publication on Northport's early history entitled *Cotton Warehouses, Saloons, and Killings*. Boyd describes four key factors that shaped early Northport. These factors shaped Northport's physical and economic landscape and were major influences in what is seen today in our downtown.

First, most of early Northport was located on a square mile of land in Township 21S Range 10W Section 16. Though an act of the US Congress passed March 2, 1819, Section 16 in every township in the state of Alabama was set aside for the use of schools. While Section 16 lands could not be sold, they could be leased. Early Northport developed under lease contracts until the legislature allowed for the selling of Section 16 land in 1827. Soon after, early Northport settler Robert Cook surveyed Section 16 and lots were sold once the lease contracts expired.

Second, the Byler and Crabbe Roads were completed around 1822. These roads allowed for goods such as cotton to be delivered to Northport warehouses and then loaded onto steamships headed for Mobile and beyond. The Byler Road is now known as Highway 43 (Main Avenue). The Crabbe Road is now known as Highway 69.

Third, the state capital was relocated to Tuscaloosa in 1826. Northport benefitted greatly from the increase in population, commerce, and government business. Transportation increased on the newly constructed Byler and Crabbe Roads and allowed for growth on the north side of the Black Warrior River.

Fourth and last, Otis Dyer was a key figure in the development of early Northport. Dyer's Ferry was one of three early ferry services that provided passage across the river. Captain Dyer, along with Reuben Dodson, surveyed the streets and lots in Northport—their work became known as the Dodson & Dyer Survey. The building of the first bridge across the Black Warrior River in 1835 eventually put Dyer's ferry service out of business.

While Northport has changed through progress, disasters, and revitalization, its landscape is much the same as it was in the days of early settlement.

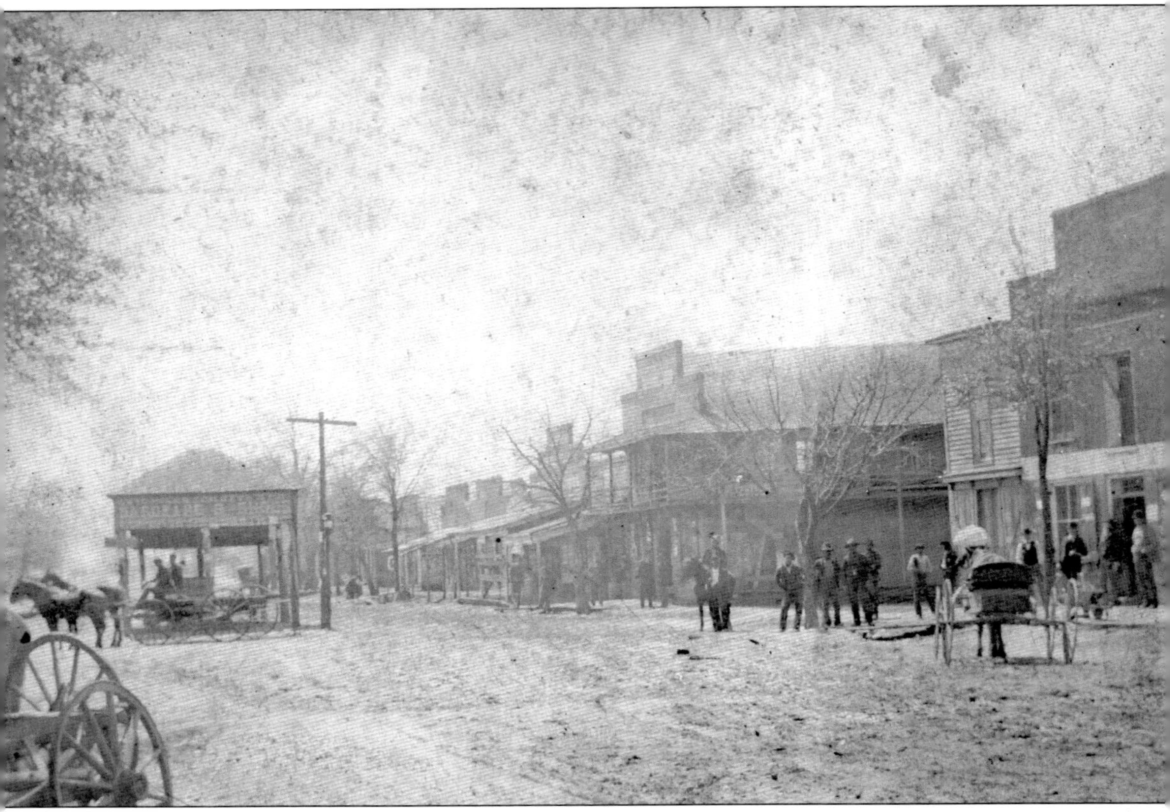

This iconic 1885 photograph of downtown Northport looks south toward the intersection of Columbus Street (now Fifth Street) and Main Avenue. The well and scale house in the middle of the street was used to sell hardware and shoes for horses. On the far right is the Christian & Daniel Building.

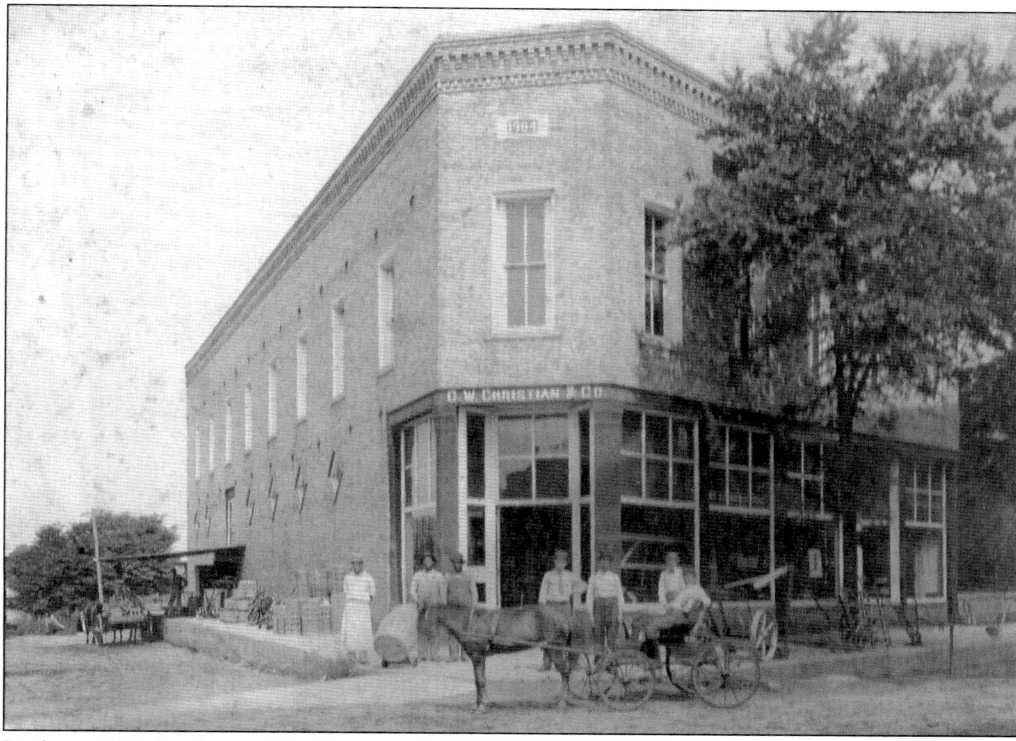

Built in 1904 for George W. Christian by W.S. Persinger, the G.W. Christian & Co. Building is on the northwest corner of Main Avenue and Columbus Street. Spiller Furniture Company later occupied the structure. In 2005, the building was purchased by Mike Chambers and renovated for commercial use.

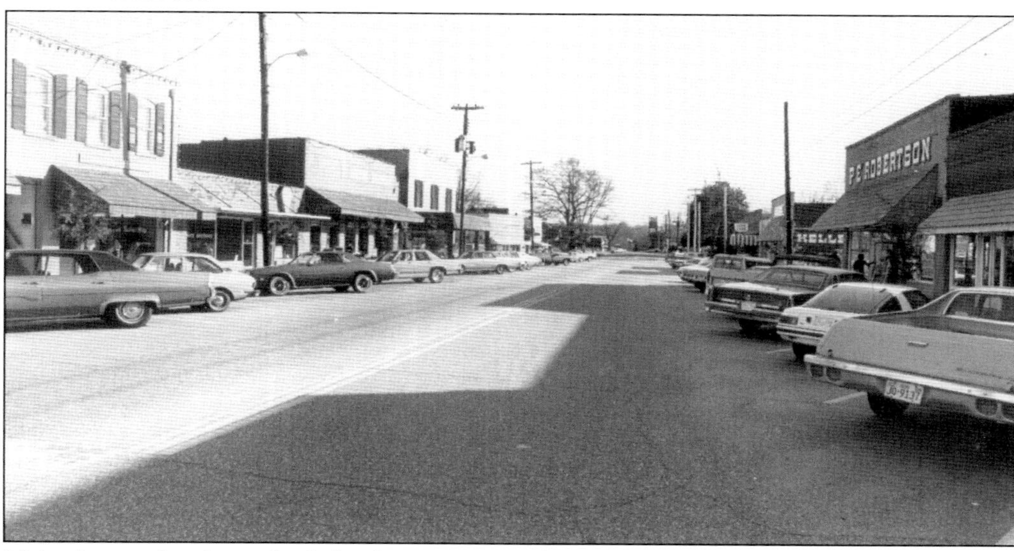

Main Avenue has been the hub of commerce in downtown Northport since the 1820s. In the mid-1990s, city leaders worked together to modernize Main Avenue and change the landscape of the downtown area. The renaissance included new lighting, streetscapes, awnings, concealed utilities, and engraved memorial brick pavers.

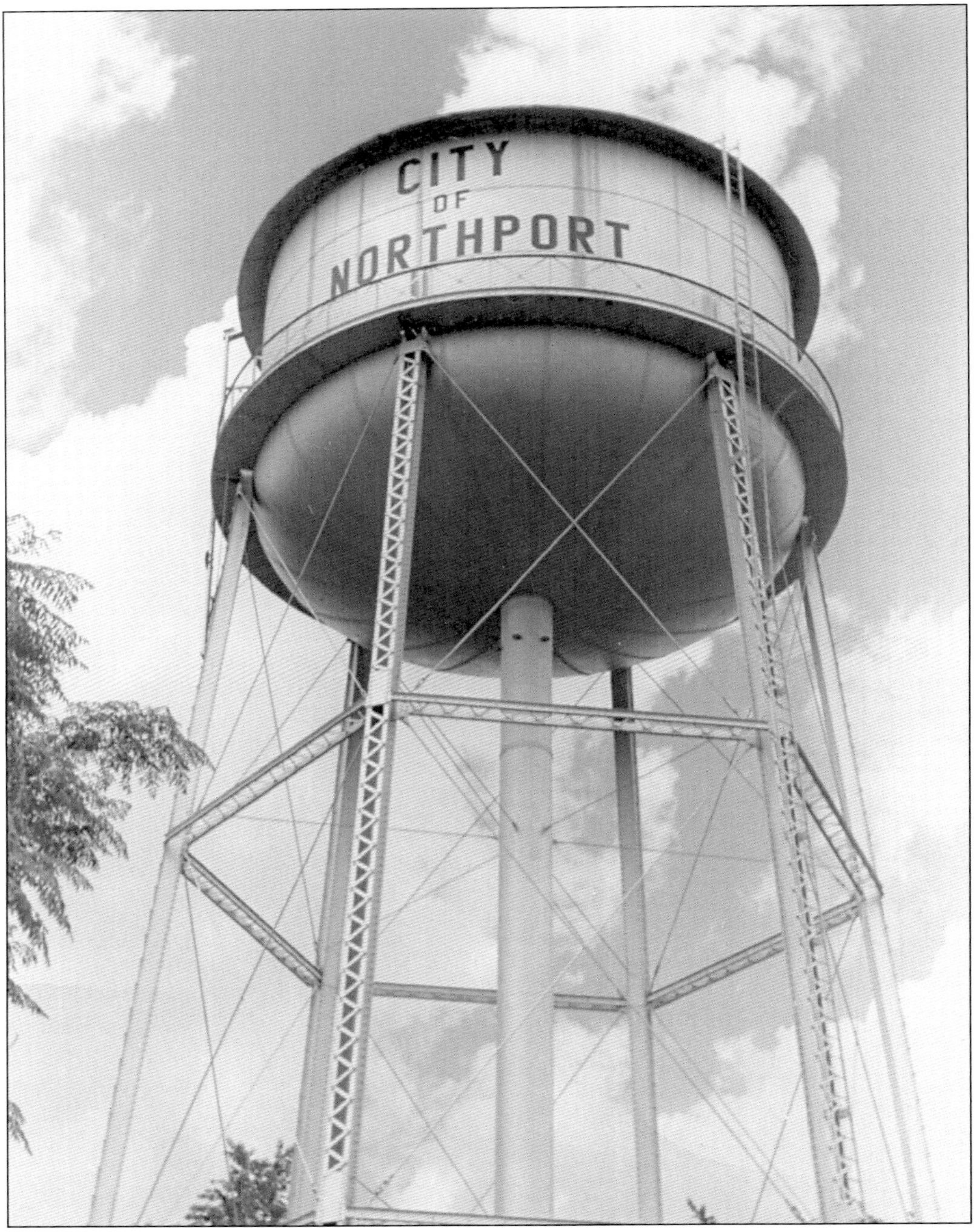

The Northport water tower stood at the foot of what is now the Twin Manor subdivision on Main Avenue. The large "City of Northport" lettering could be seen by motorists crossing the river bridge into Northport. An earlier tower made of cement still stands in the same location. This photograph was taken in the 1950s.

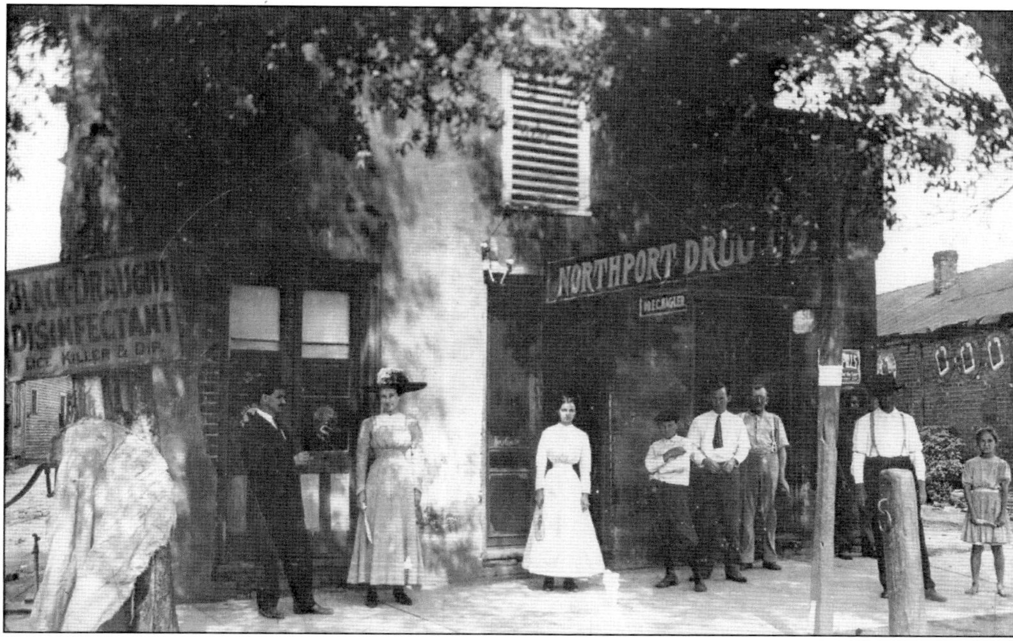

Dr. Samuel T. Hardin (far left) operated a drugstore on the west side of Main Avenue from 1898 to 1909. He established his medical office at the old Northport Drug Co. but made house calls to many patients. The building was torn down and the space is now a public parking lot next to Baby Bundles, at 420 Main Avenue.

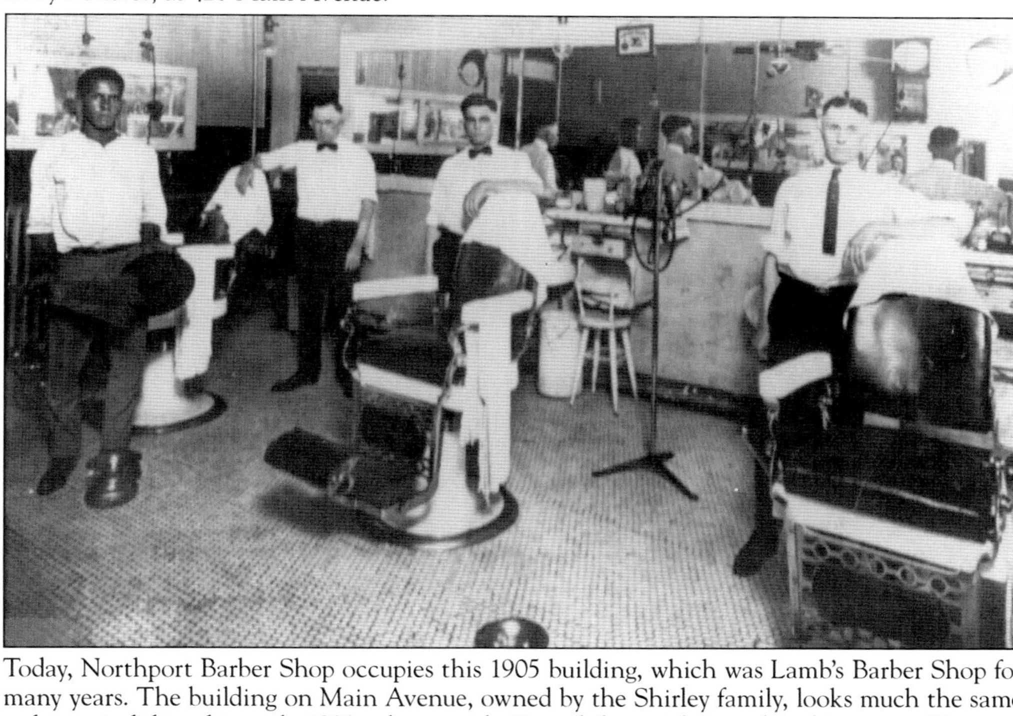
Today, Northport Barber Shop occupies this 1905 building, which was Lamb's Barber Shop for many years. The building on Main Avenue, owned by the Shirley family, looks much the same today as it did in this early 1920s photograph. From left to right are shoeshine man Clarence Ellis, Vird Palmer, B.B. Smalley, and Billy Palmer.

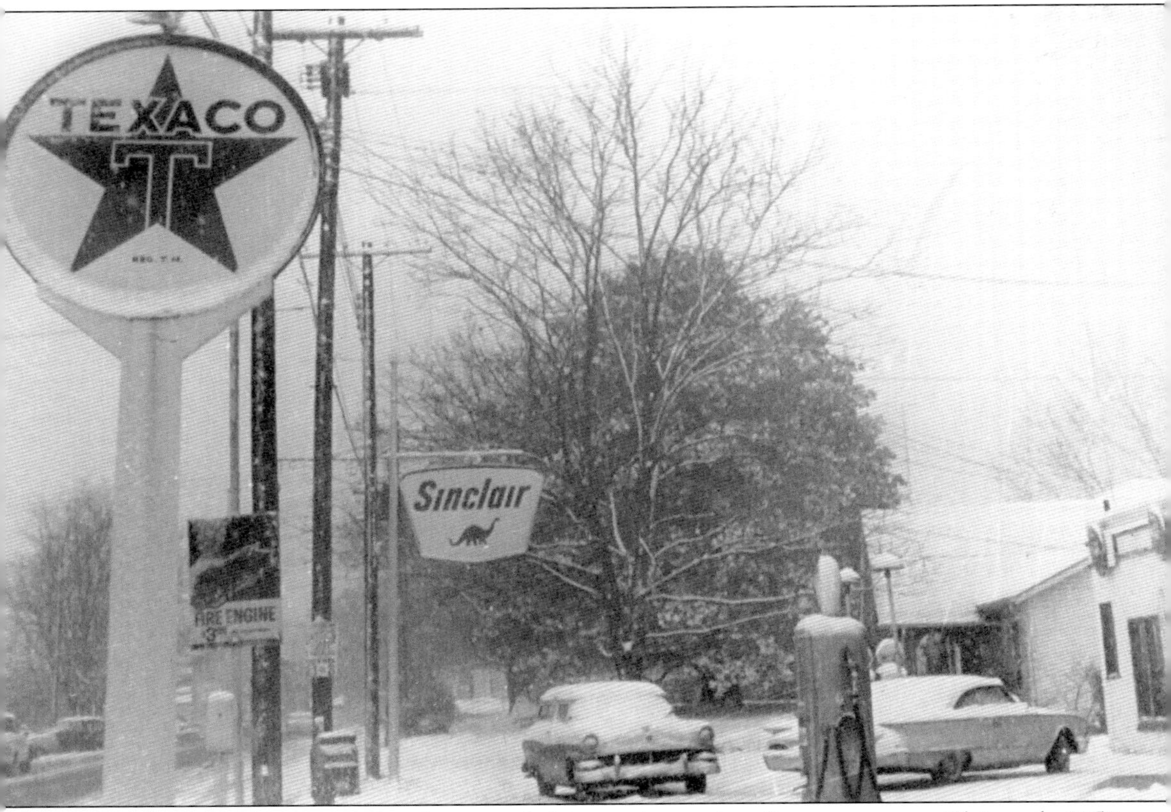

Bailey Thomson owned and operated this service station for many years. This early 1950s photograph shows an early location next to Maxwell Tire & Oil Co., where City Café is today. Thomson later moved to the northeast corner of First Street and Main Avenue, across from the cotton gin. Many locals remember the distinctive Sinclair sign with the dinosaur logo.

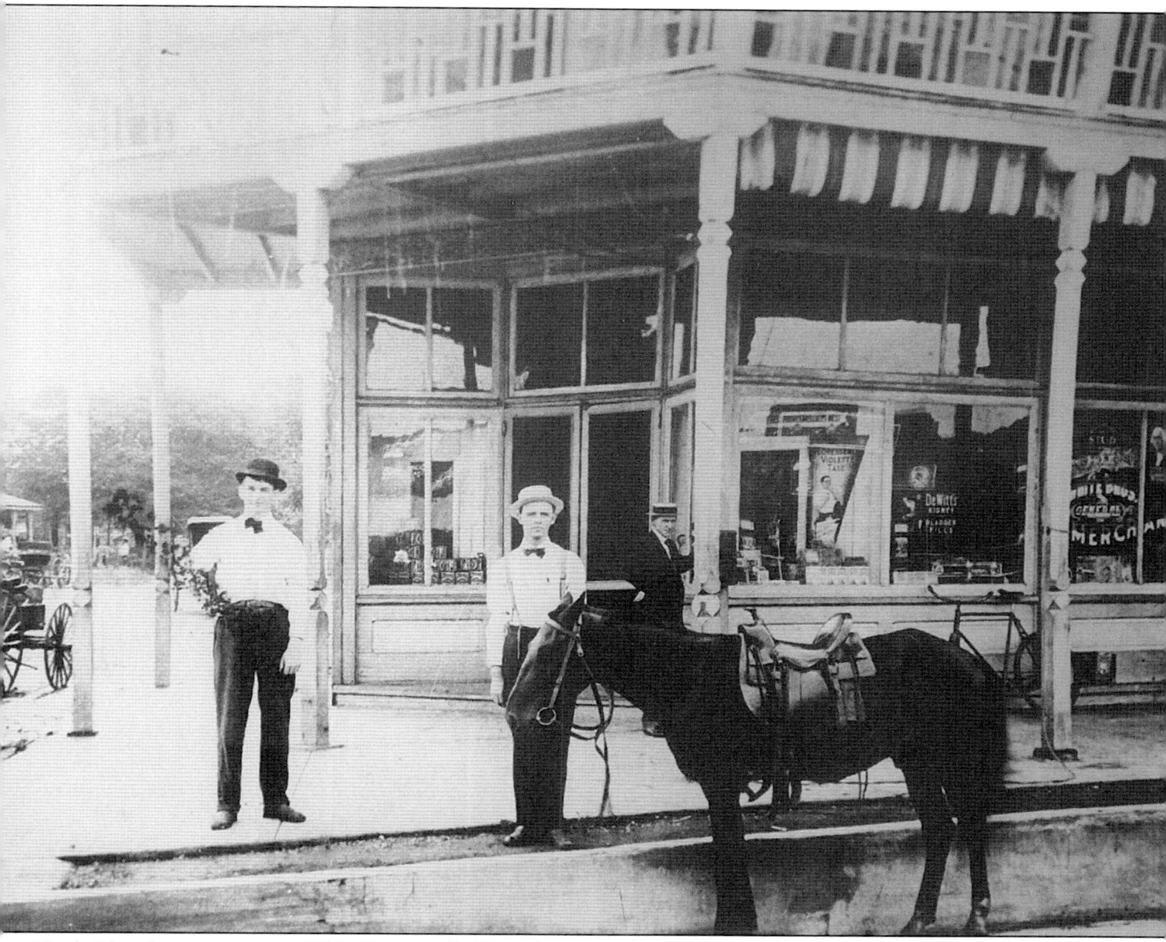

Early Northport citizens W.A. Barnes (left) and Vasa Adams (center) are seen in 1910 in front of White Bros. General Store. The structure was built in 1905 on the southeast corner of Columbus Street and Main Avenue. The tornado of 1932 demolished the second story. It is now the home of Fifth & Main Restaurant.

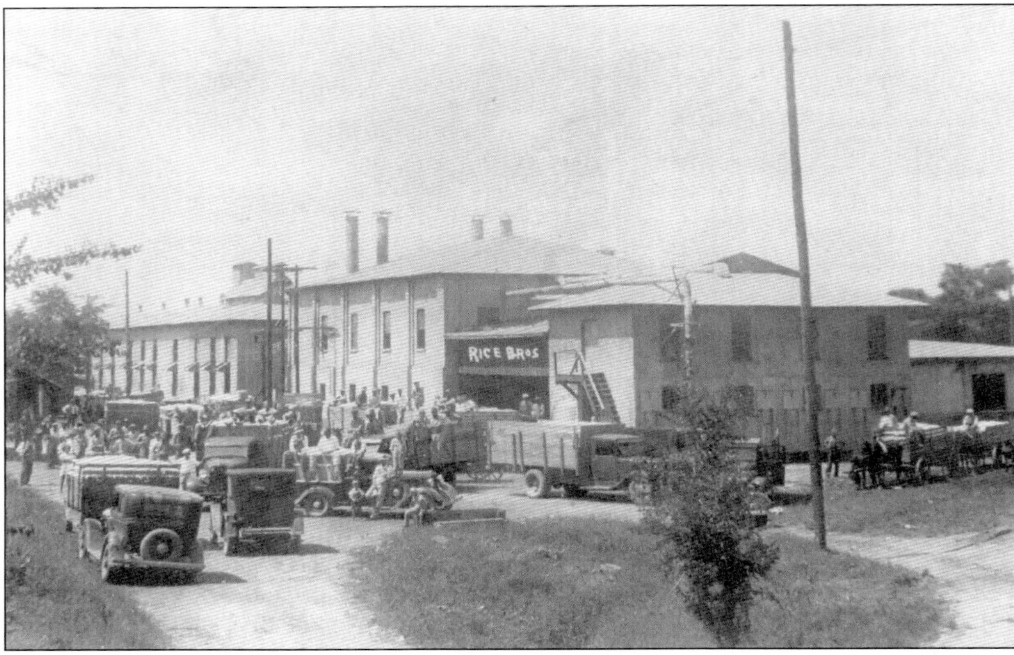

Rice Bros. Gin and Warehouse, on Tenth Street just west of Northport Baptist Church, was a booming operation for more than half a century. Brothers Frank and Joe Rice started the business around 1918 and this photograph was taken in the 1920s. Turner Warehouses occupies the site today.

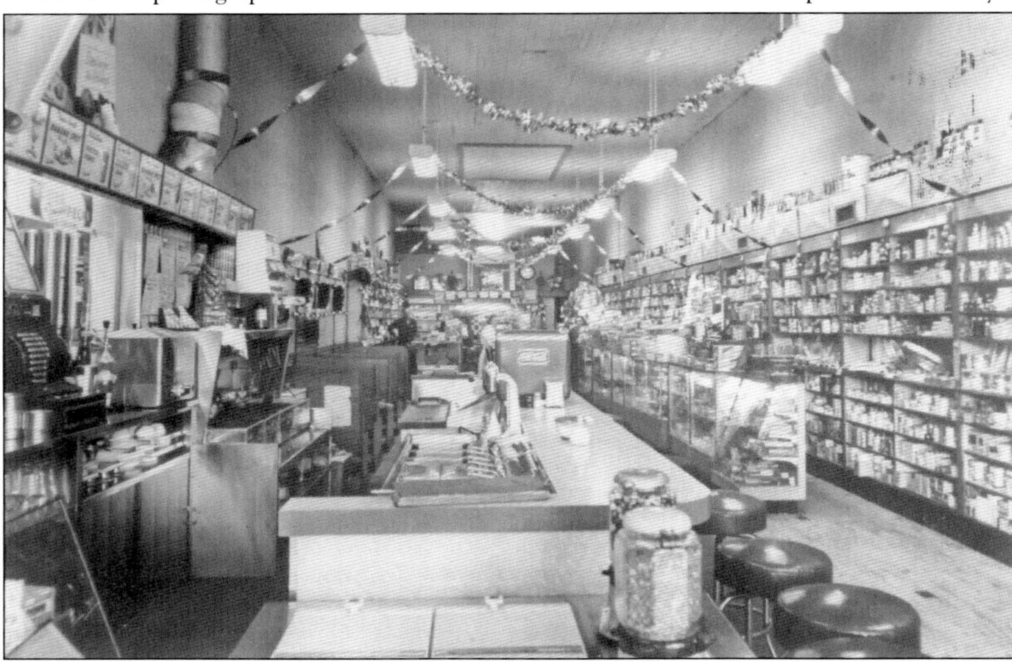

Adams Drug Store was a landmark business on the west side of Main Avenue. In the far back of the store are Carl Adams Sr. and his wife, Catherine. Today, their son, Carl Adams Jr., operates Adams Antiques from the same location, continuing the family tradition that has been going for 75 years.

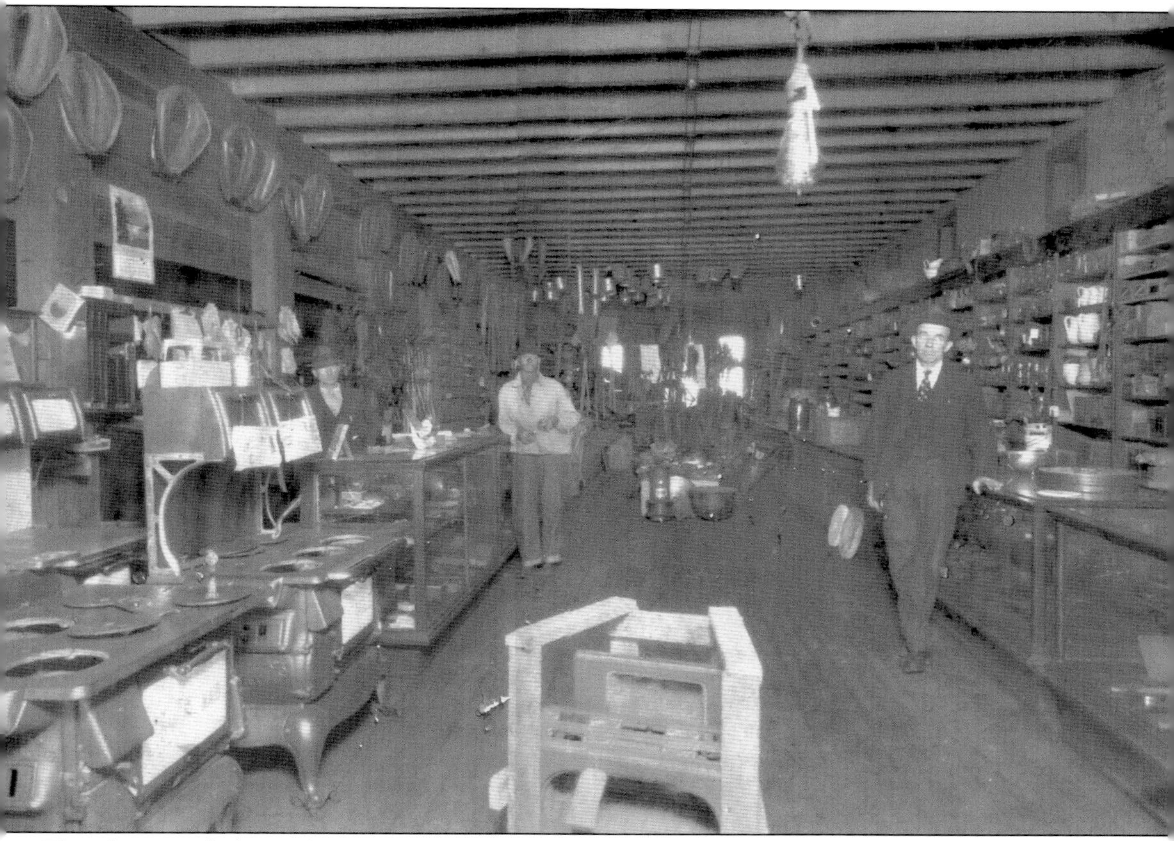

This photograph shows the inside of Anders Hardware in the 1950s. Many locals will recognize the doors that lead to the sidewalk along Main Avenue. The business, started in 1909, was originally called J.H. Anders Mercantile Co. and later Northport Furniture and Hardware Company. On August 7, 1930, the brothers incorporated the family business and renamed it Anders Hardware Company.

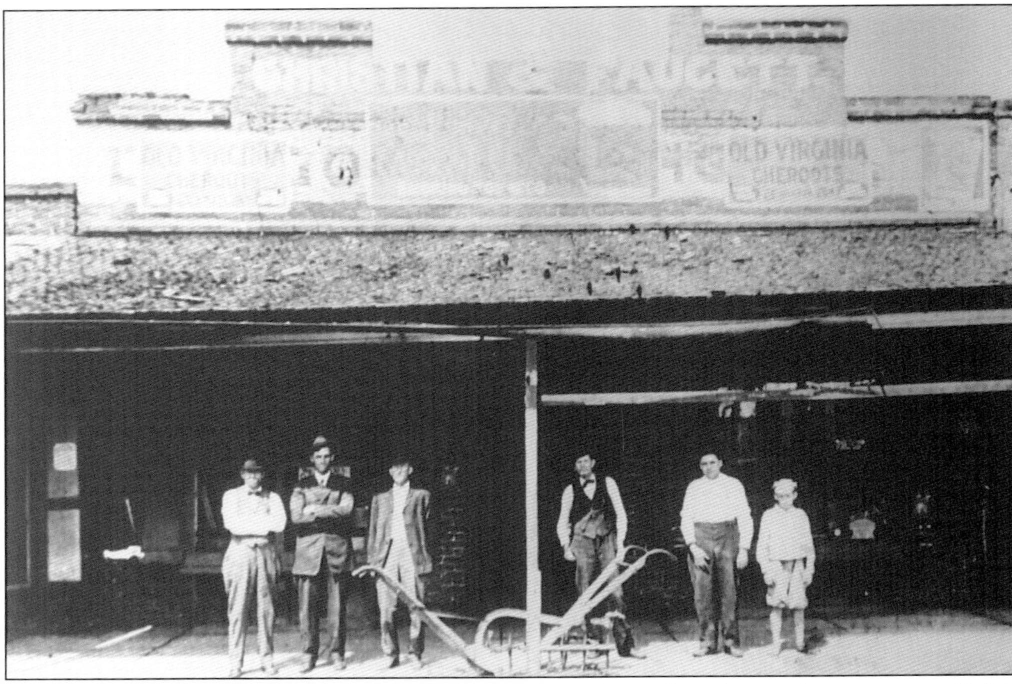

Christian & Faucett, a general mercantile business, was located on the northwest corner of Main Avenue and Columbus Street. The structure seen here was replaced by the 1904 G.W. Christian & Co. Building. This late 1800s photograph shows, in no particular order, S.P. Faucett, Jake Anders, Clint Deason, and Jim Adams (directly right of the post).

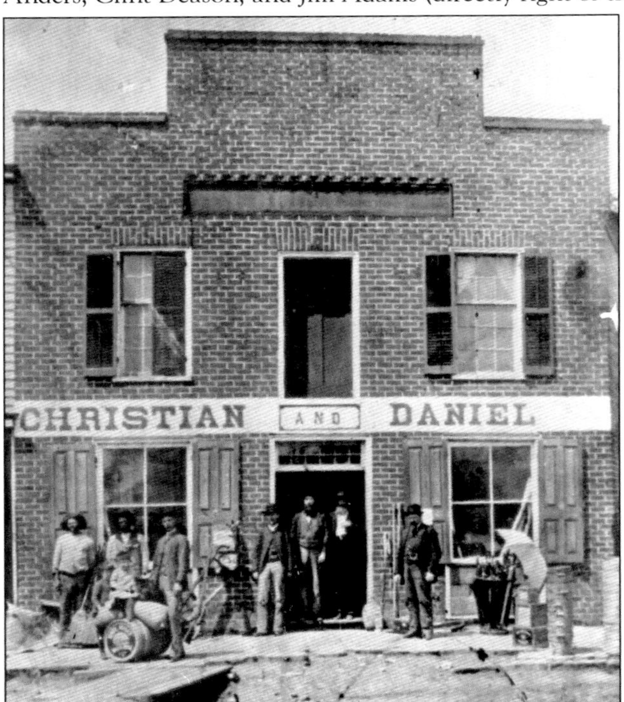

Formerly known as Shirley & Christian and later as Christian & Faucett, the Christian & Daniel General Mercantile was operated by William Lewis Christian and his son-in-law, John M. Daniel, who had married Christian's daughter Amanda. The structure, built by Christian's brother-in-law, James Shirley, is notable for the parapet extending above its roofline. This photograph was taken in the late 1800s.

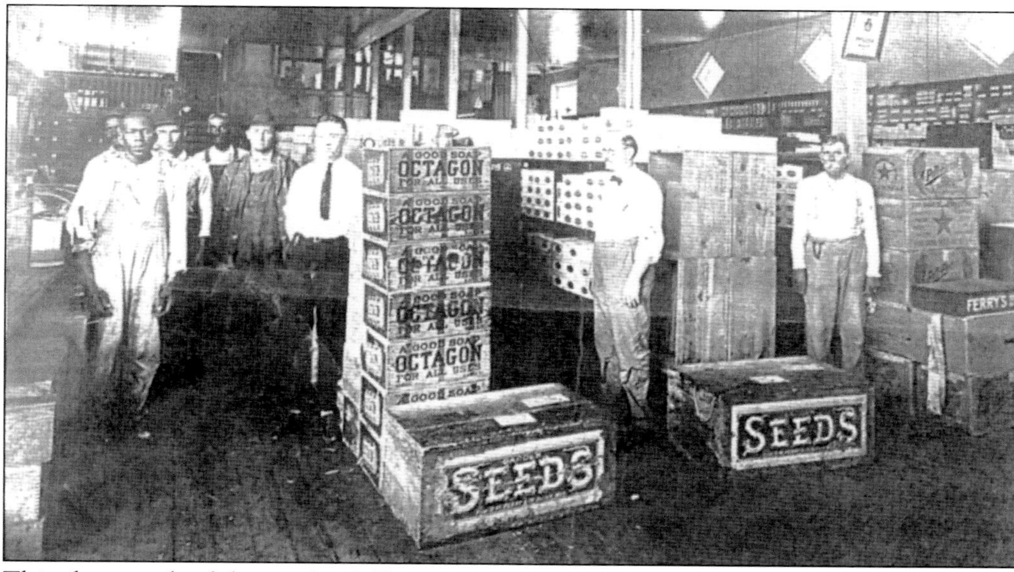

This photograph of the inside of Christian & Faucett shows Ollie Ramsey (on left with white shirt and tie), Sam Palmer (with bow tie and glasses), and Elijah Moore (far right). This structure, now known as the G.W. Christian & Co. Building, was purchased by Mike Chambers in 2005 and restored for commercial and office space.

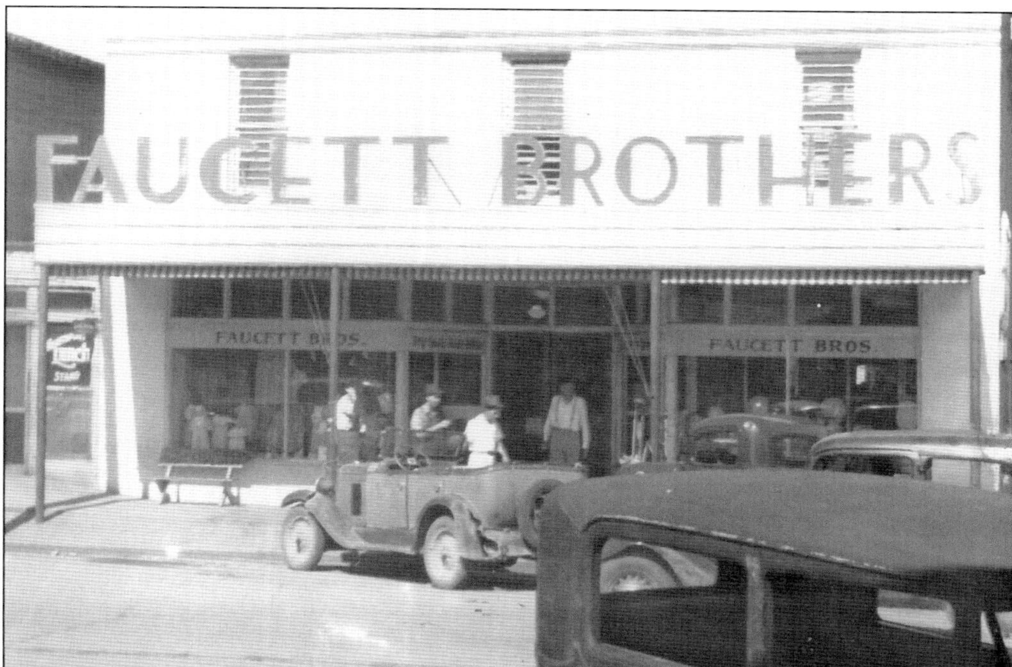

Located on the east side of Main Avenue just south of Anders Hardware, this structure was originally built as a post office for Northport. It later housed the Northport Bank. In 1937, the Faucett general mercantile located here, later becoming Faucett's Department Store. In the enclosed alley next to the store was a small diner. An early advertisement for the business read "Faucett Brothers Merchants-Cotton Buyers-Fertilizer Manufacturers-Phone 1407 Northport, Alabama."

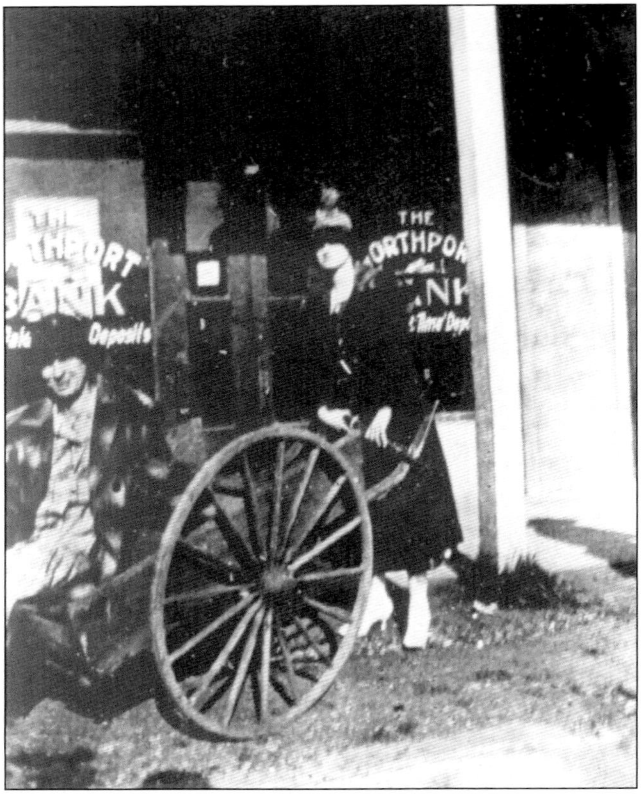

The Northport Bank was the city's first bank. The operation opened in late 1913 with S.A. Yerby as president and closed in October 1933. In this photograph from around 1920, Matt Lindsey is sitting in the cart, which was used by Ninnie Cummins to haul mail from the Northport railroad depot to town.

S.M. Freeman & Son General Mercantile (below) was located in the 1850 Shirley Building on the west side of Main Avenue. Next door, there was a business owned by U.S. Lindsey. S.M. Freeman is seen here, second from right, with family members in the 1890s. Now the home of Northport 5 & 10, it is the oldest surviving commercial structure in Northport.

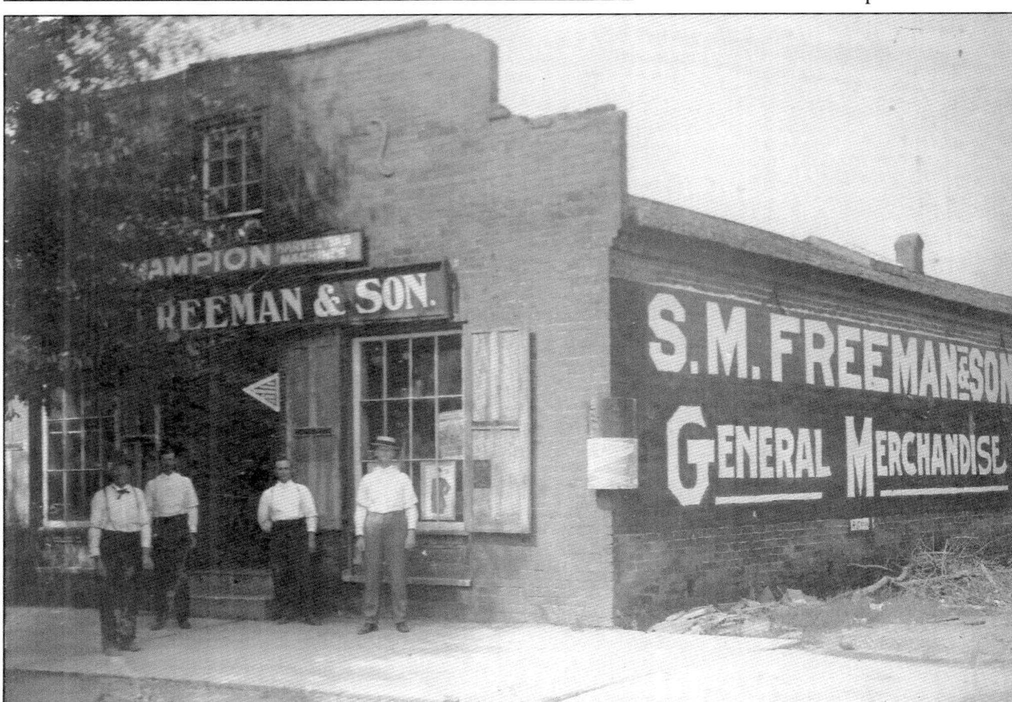

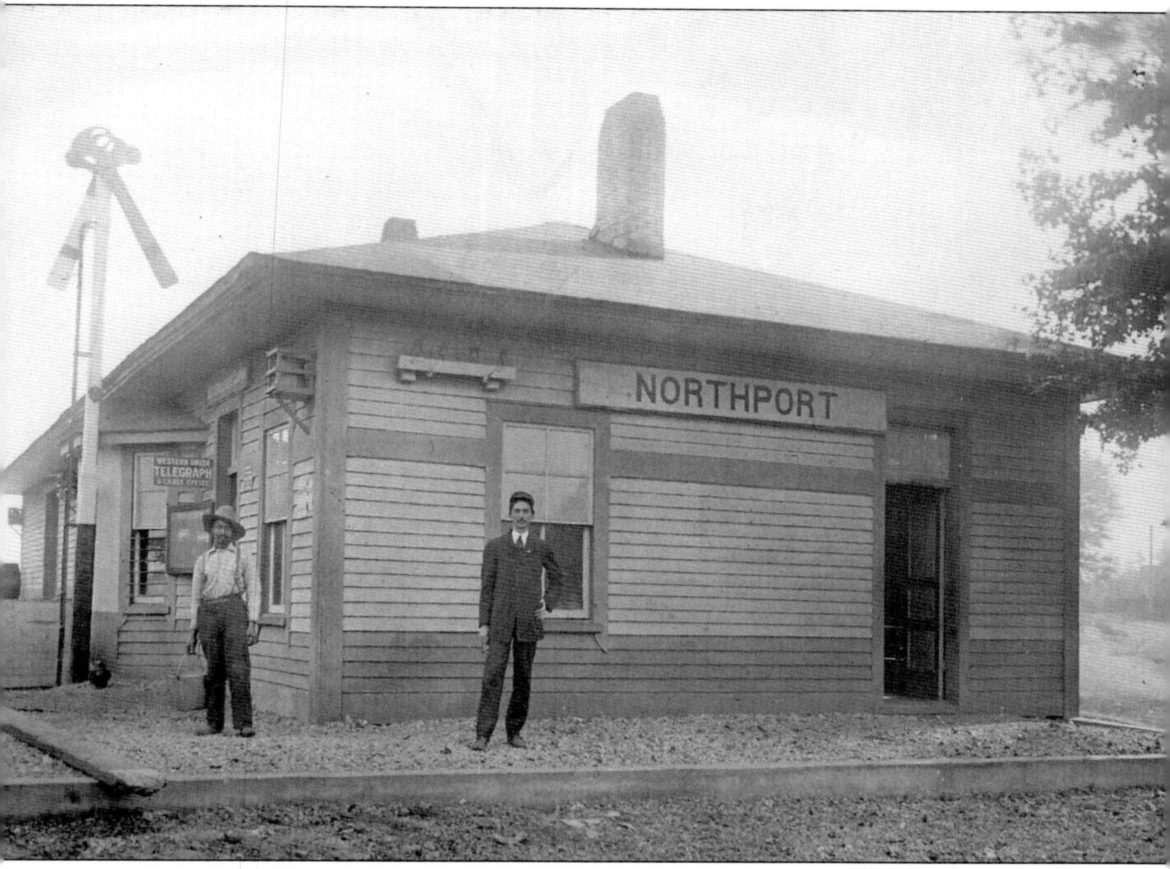

Ninnie Cummins (left) and Richard Pearson are seen here in front of the Northport railroad depot in 1902. Cummins delivered mail to the Northport post office after it arrived on the Gulf, Mobile & Ohio train. The depot was built in 1897 and burned down on Christmas Eve 1923. It was rebuilt in 1927.

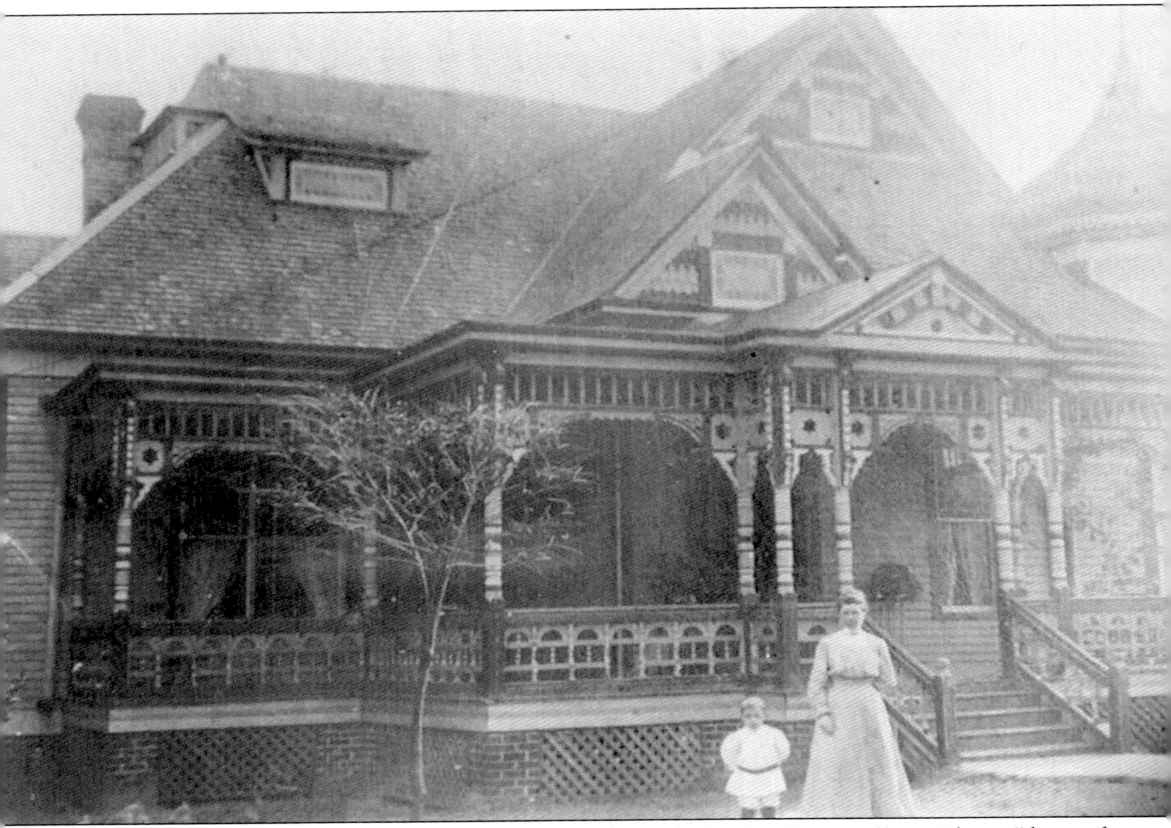

This Queen Anne–style home was built around 1890 by Richard Maxwell as a "dowry" house for his daughter, Clara Maxwell Hamner. The Maxwell-Hamner House, known as the "Gingerbread House" by many, is now occupied by Main Avenue Antiques. Clara Maxwell Hamner is seen here with one of her children.

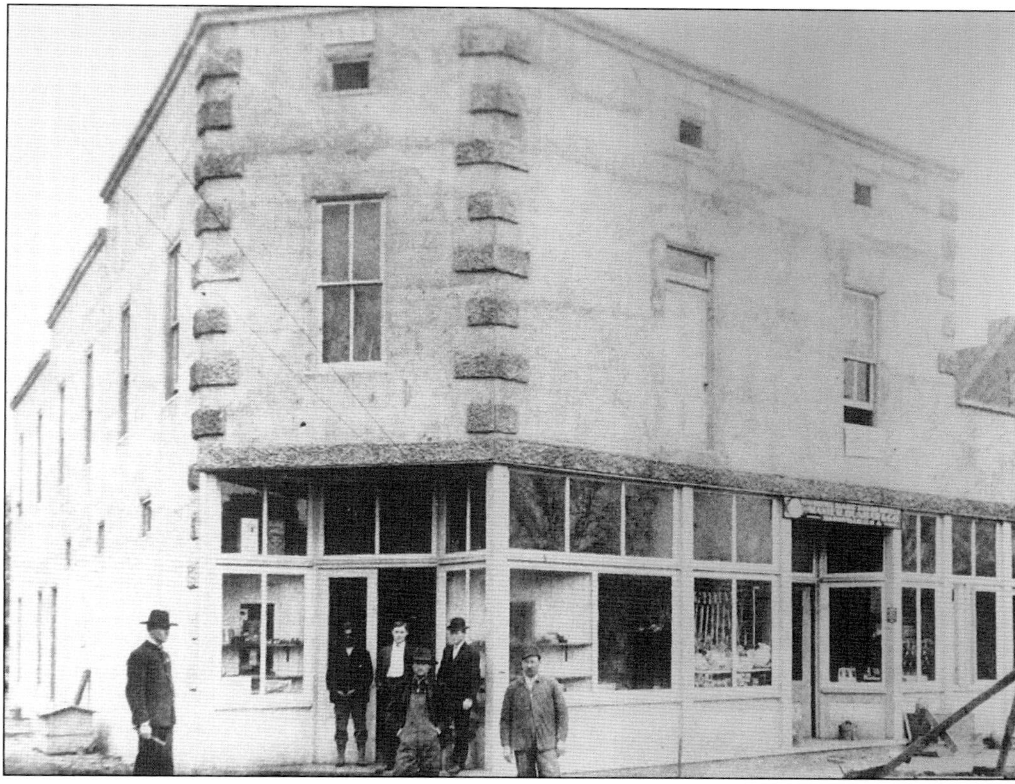

This is the earliest known photograph of White Bros. General Store. The structure was built in 1905 on the southeast corner of Columbus Street and Main Avenue. Northport resident Burton Hyche is standing on the far left.

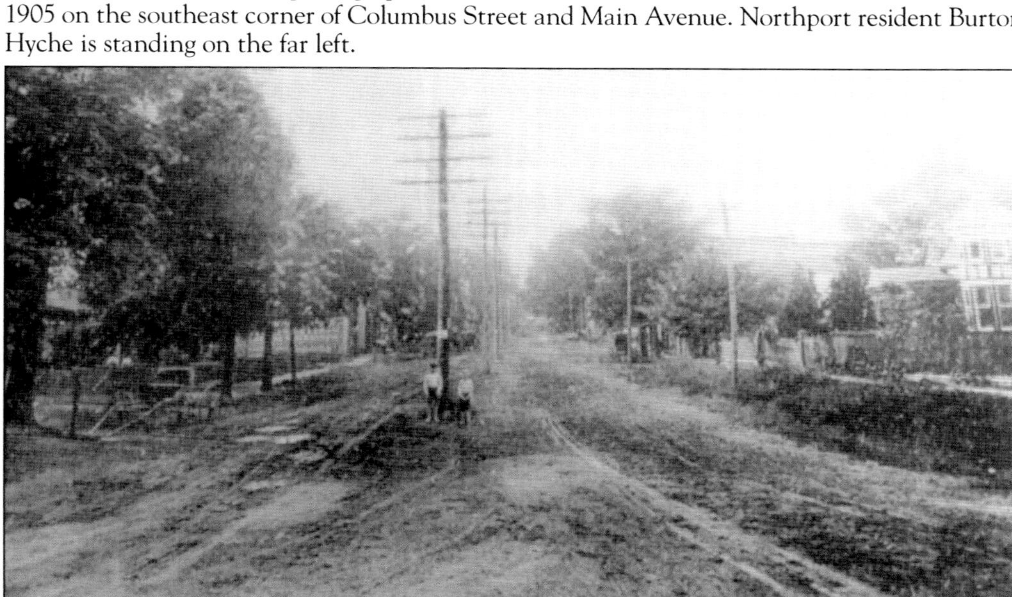

This early 1900s photograph shows the 1867 Maxwell-Peters home on Main Avenue. The Maxwell-Hamner "Gingerbread House" is on the right. Telegraph poles lined the dusty downtown streets during this time. Two barefooted young children are in the center of the street.

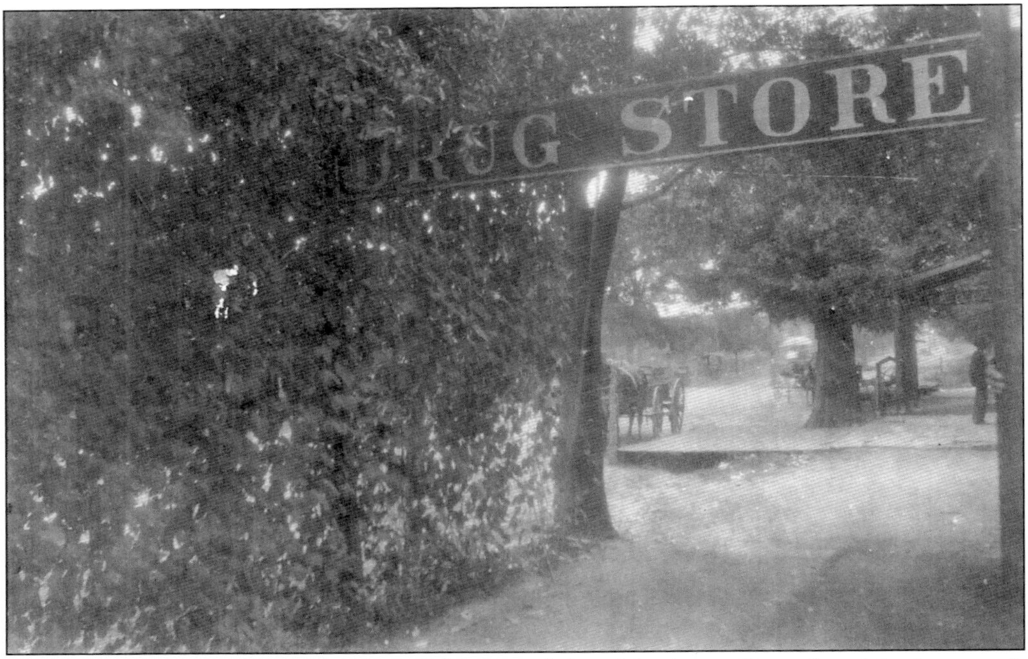

Looking down Columbus Street (now Fifth Street) from Northport Drug Co., this photograph shows the west side of Main Avenue. Dr. B.F. Powell established his medical practice here around 1838 and was followed by Dr. Samuel T. Hardin, who practiced here from 1898 until 1909, and Dr. E.C. Hagler.

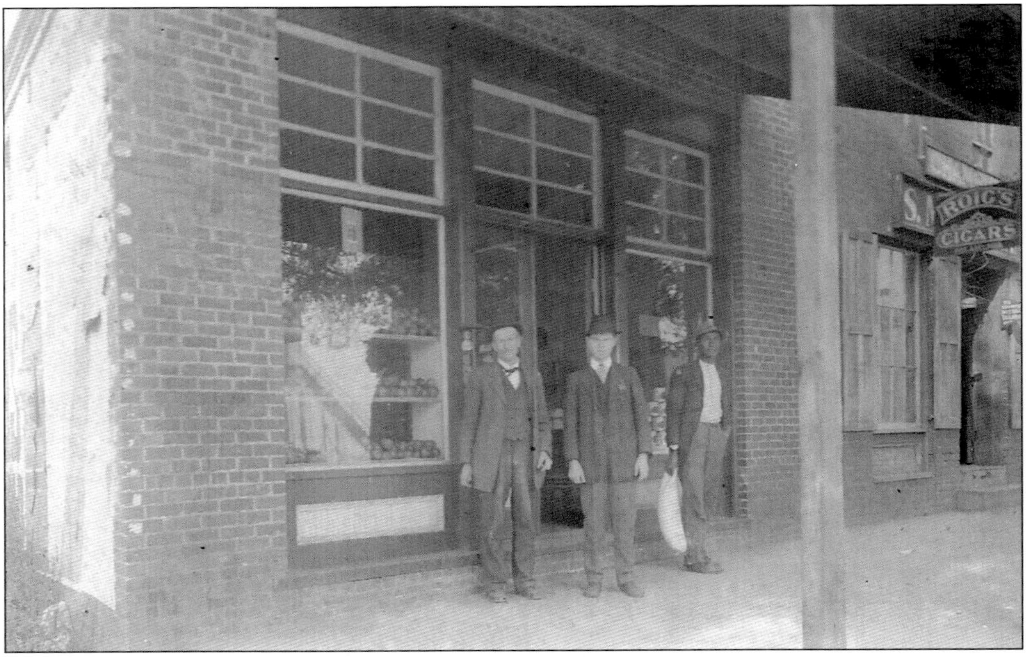

Proprietor U.S. Lindsey (center) stands with two unidentified men in front of the Lindsey Store on the west side of Main Avenue around 1915. Eyecatchers Screen Printing currently occupies the building.

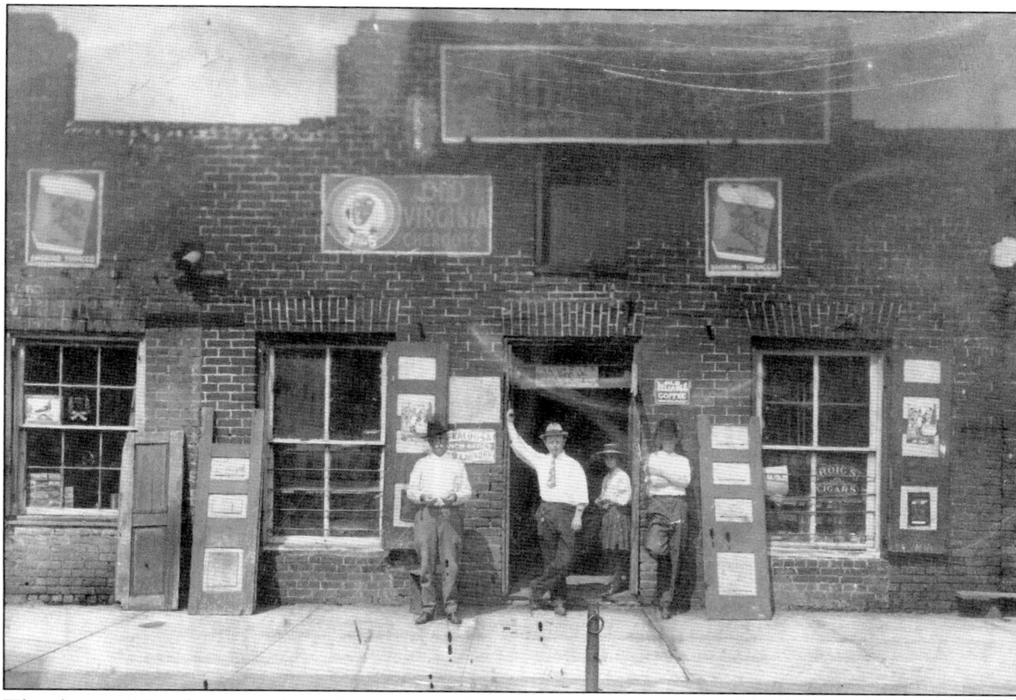

This late 1800s photograph of downtown Northport shows an early building on the west side of Main Avenue, where Adams Antiques is today. It was built in the design that James Shirley made popular with other 19th-century commercial structures.

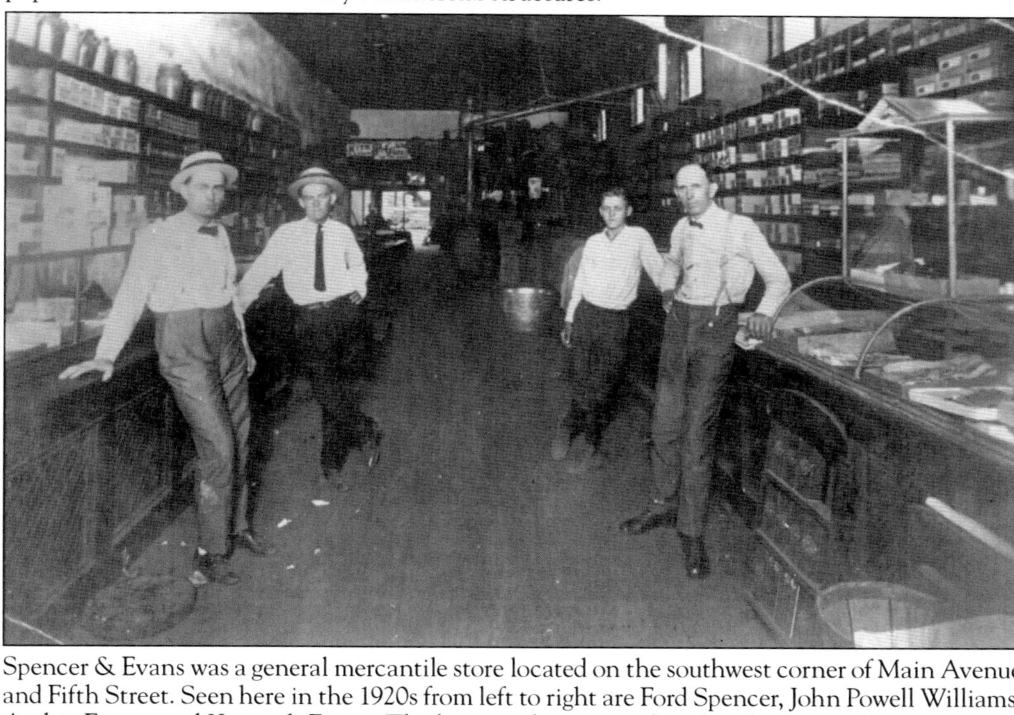

Spencer & Evans was a general mercantile store located on the southwest corner of Main Avenue and Fifth Street. Seen here in the 1920s from left to right are Ford Spencer, John Powell Williams, Archie Evans, and Kenneth Evans. The business later moved to the Shirley Building.

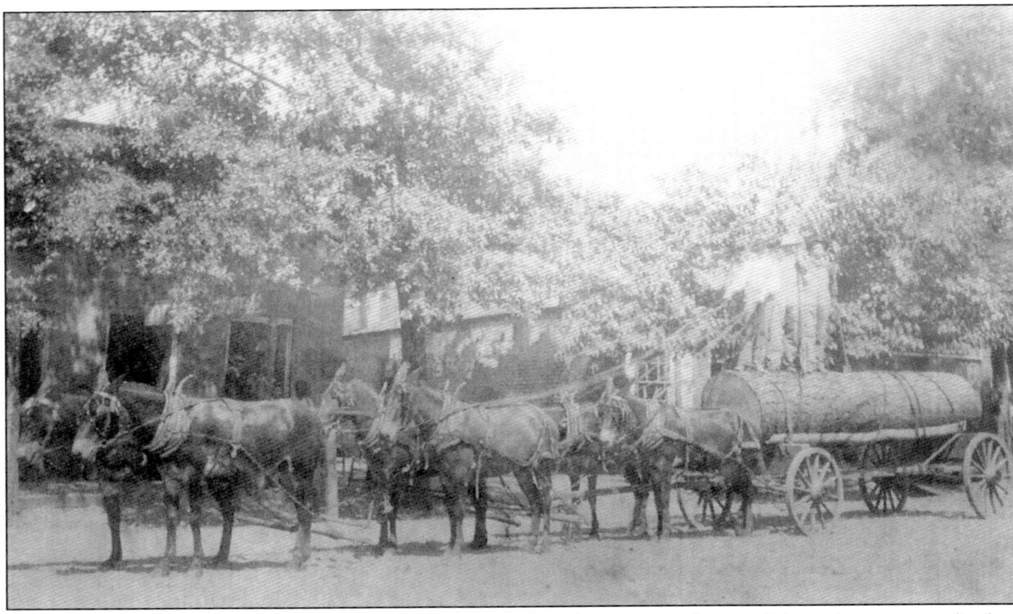

A six-mule team wagon operated by the Rushing family carries a log on Main Avenue in the late 1800s. This was a familiar sight in Northport, as timber was brought to market and cut in one of several downtown sawmills. Note the size of the log and the length of the wagon carrying it. (Courtesy of Heritage Commission of Tuscaloosa County.)

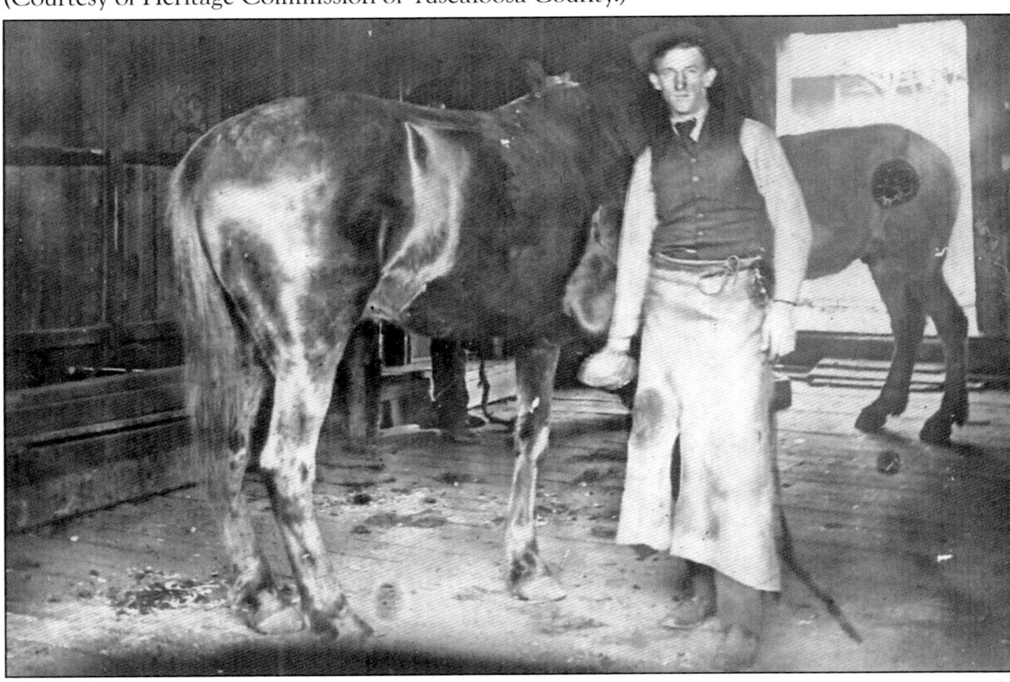

Josh Palmer is seen here in his blacksmith shop on Columbus Street. In 1899, as a young man, he purchased property in Northport and established his shop, which he operated for more than 45 years. The shop was located where a part of the Kentuck Museum complex is today. Palmer made and repaired horseshoes for a majority of the horses in Northport until the late 1940s.

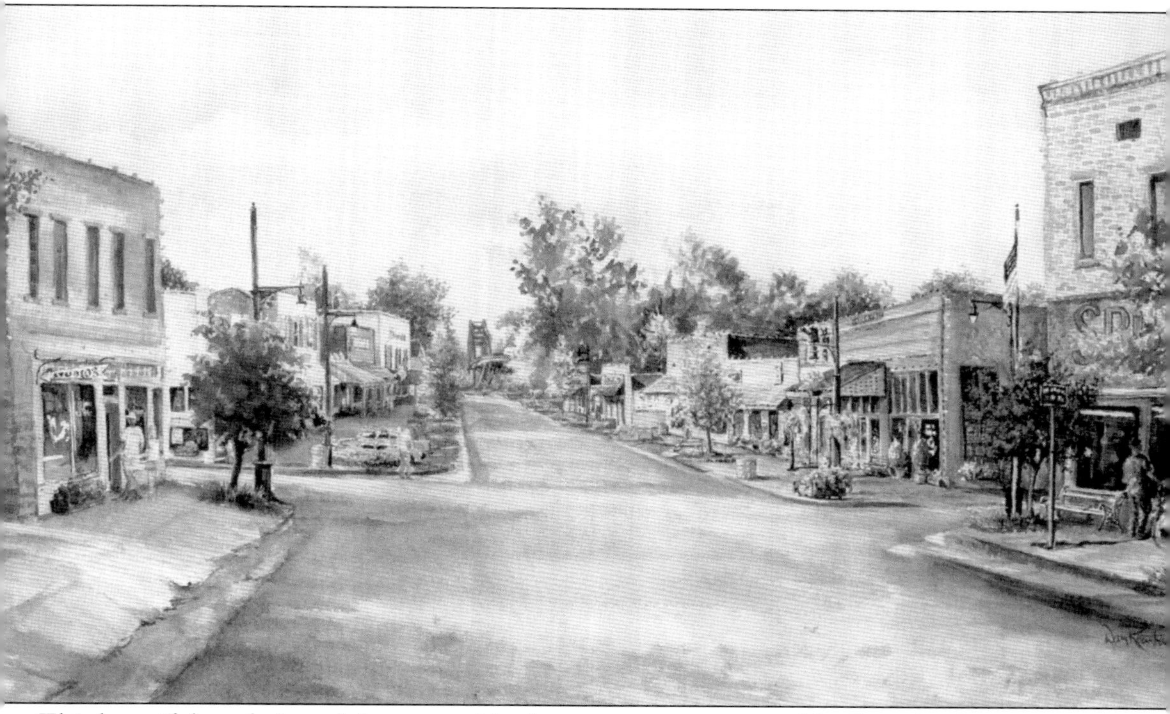

This beautiful rendering of downtown Northport was commissioned by Friends of Historic Northport and painted by local artist and noted watercolorist Danny Rountree. Prints from the original Rountree work continue to sell years after the original work was completed.

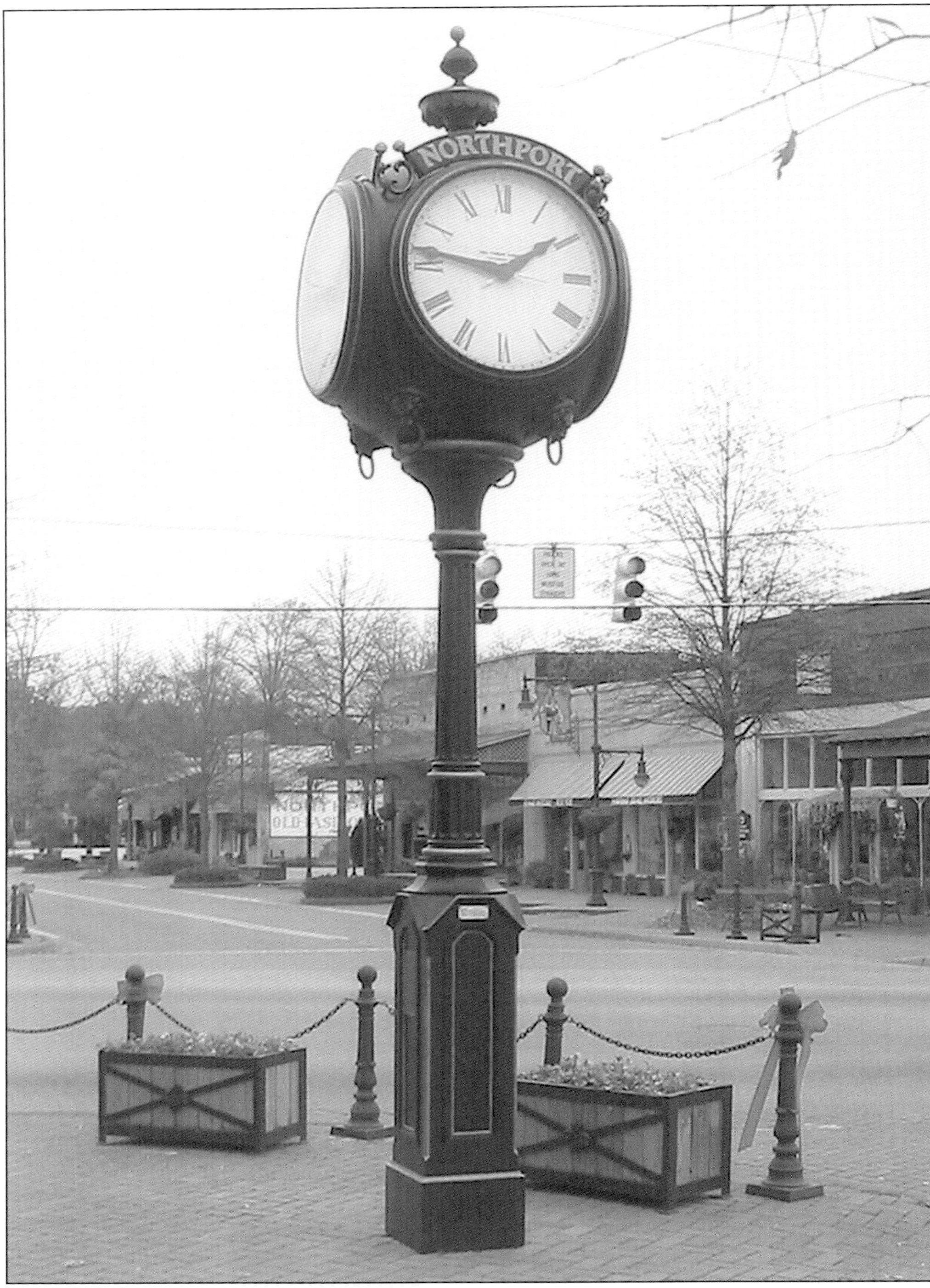

Donated by local developer Stan Pate in 1999, the Northport Town Clock sits proudly in front of the Kentuck Museum on Main Avenue. Its beautiful chimes and music are heard by those walking or driving along Main Avenue, and it has greatly enhanced interest and investment in historic downtown Northport.

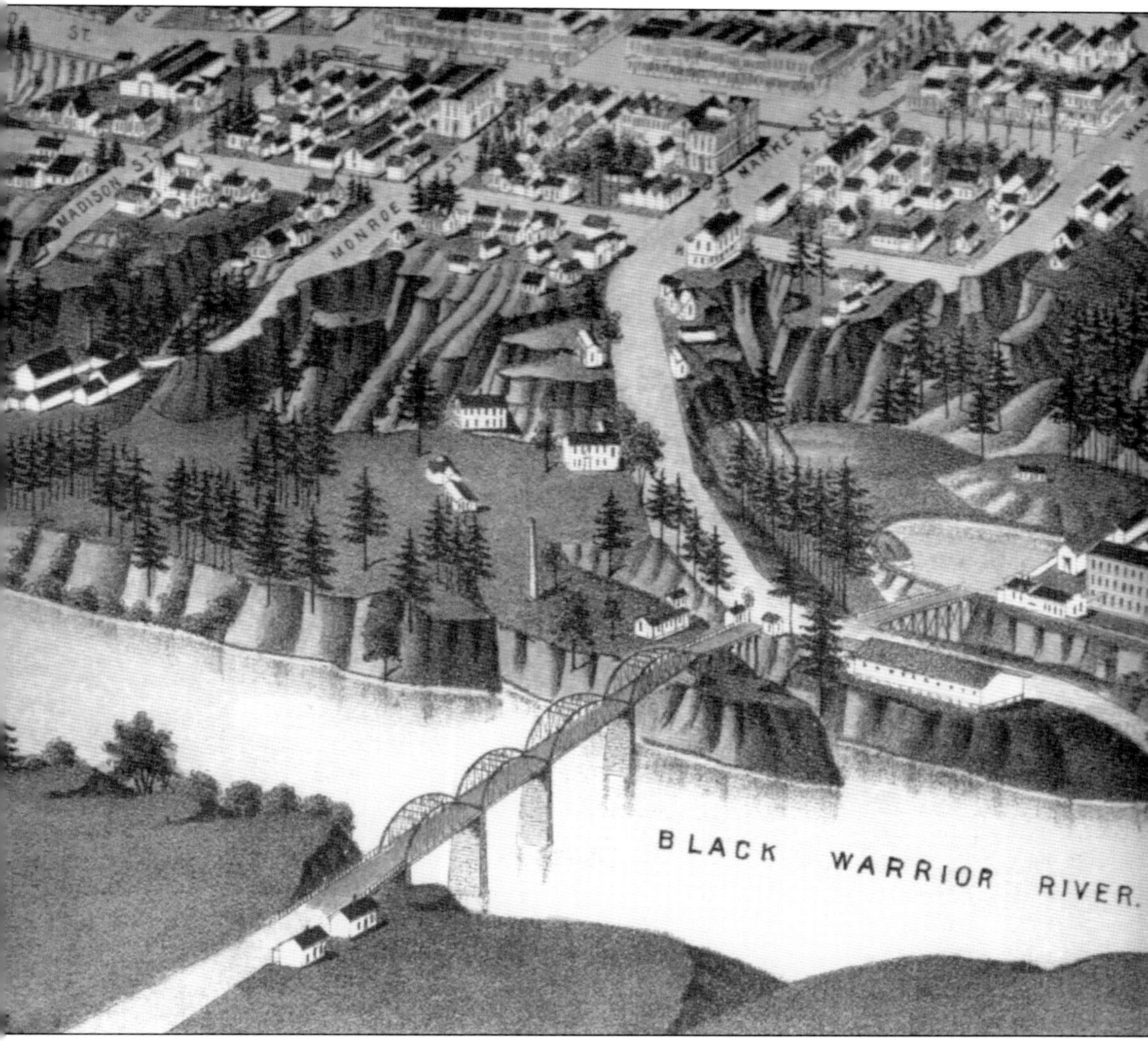

This section of an 1887 perspective map of Tuskaloosa by Henry Wellge shows the 1882 King Bowstring Bridge crossing into Northport. The cliffs at the foot of the Black Warrior River allowed Tuscaloosa to avoid flooding problems. Northport had to contend with flooding until the Richard L. Platt Levee was completed in the late 1990s. (Courtesy of Library of Congress, Geography and Map Division.)

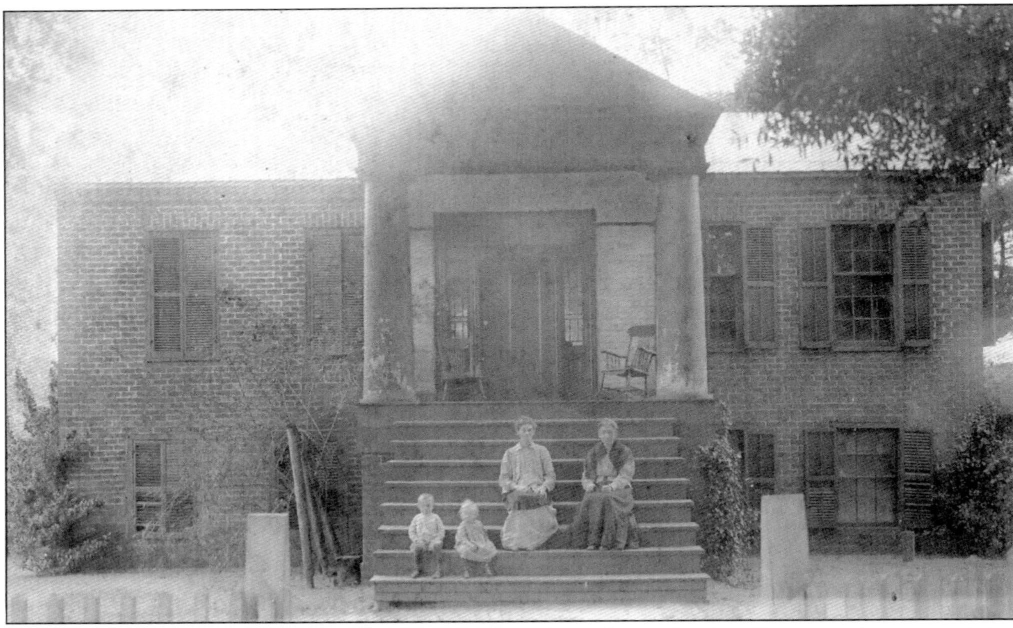

This c. 1900 photograph shows historic Shirley Place, formerly the Christian home, on Main Avenue in downtown Northport. The house was built in 1838 by James Shirley, an early Northport merchant. It was built of handmade brick in a Raised Cottage style using Federal and Greek Revival detailing. Two massive Doric columns support a portico porch roof over the main entrance. Sitting on the front steps are members of the H. Clinton Deason family.

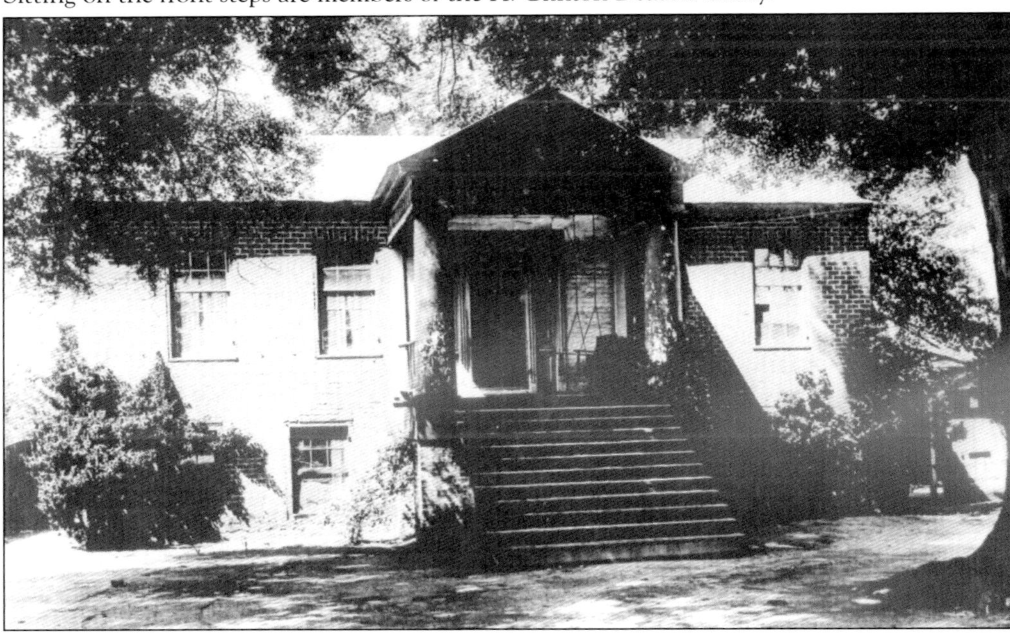

This photograph shows historic Shirley Place in the 1950s. After Marvin Harper purchased the home in 1979, he gave it the name Shirley Place in honor of its builder and first occupant, James Shirley. It is the oldest building in downtown Northport and features four exterior chimneys that serve eight fireplaces.

Two

Disaster Strikes

The tornado that struck Tuscaloosa on April 27, 2011, is the most devastating disaster to strike the greater Tuscaloosa community. The three-day outbreak of tornadoes from April 25 to April 28 was the largest outbreak of tornadoes on record.

While the 2011 tornado did not hit Northport, it killed more than 50 members of the community and many more around the state. Northport sent its policemen, firemen, and public works employees to Tuscaloosa to aid in the recovery efforts and the effects of this tragedy were felt deeply in Northport.

Few other tornadoes can begin to compare with the one experienced on April 27. However, there is a history of tornadoes in the area. On March 4, 1842, a twister leveled Newtown, a community west of Tuscaloosa now known as West End. The tornado rendered the covered bridge crossing the Black Warrior River unusable, moving it two feet upriver and blowing off much of its roof.

On March 21, 1932, an F4 tornado struck downtown Northport. A clock at the demolished Tuscaloosa Country Club stopped at 4:01 p.m., about 30 minutes after the tornado first struck near Demopolis. After touching down at the western end of Tuscaloosa, the tornado moved across the Black Warrior River into Northport. Witnesses said that it was shaped like an ice-cream cone when it moved through the heart of Northport, and that it was so filled with debris that it had an eerie white glow and resembled a heavy snow shower moving in on the city.

The 1932 tornado killed 38 people and injured 250 more in Northport. Druid City Hospital in Tuscaloosa was quickly filled to capacity. The University of Alabama gymnasium was also pressed into service as an emergency additional hospital. The *Tuscaloosa News* reported, "It looked as if Northport had been bombed."

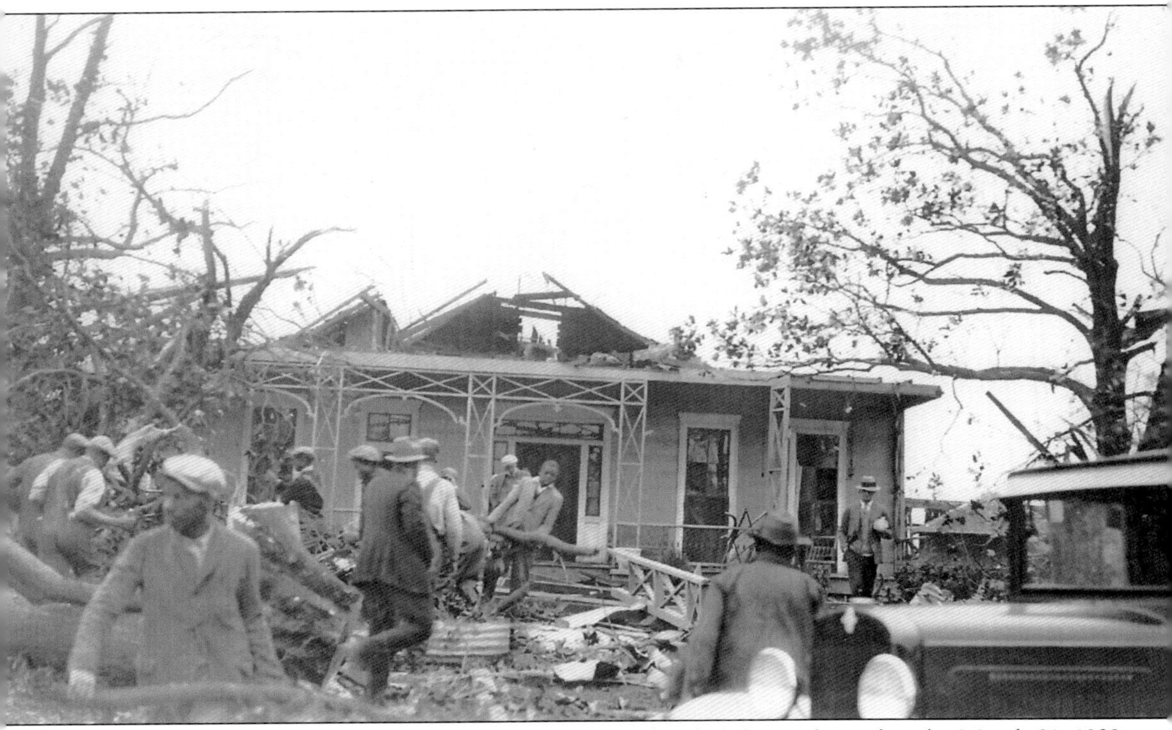

The Maxwell-Peters home is seen here after it was barely left standing after the March 21, 1932, tornado. The 1867 structure, built in the English Cottage Orné style, was owned by Dr. William Marcus Peters and his wife, Pearl Maxwell Peters. While the home was damaged, the two magnolia trees planted in 1868 by Richard Maxwell survived. After the storm, Dr. Peters set up a triage area at the rear of the house to care for victims of the tornado.

This photograph shows a Fourth Street and Main Avenue street sign after the flood of 1979. The Northport City Shop, on Fourth Street one block from Main Avenue, had to be completely evacuated. Reaching the 65-foot mark, the 1979 flood is considered the last major flood in downtown Northport.

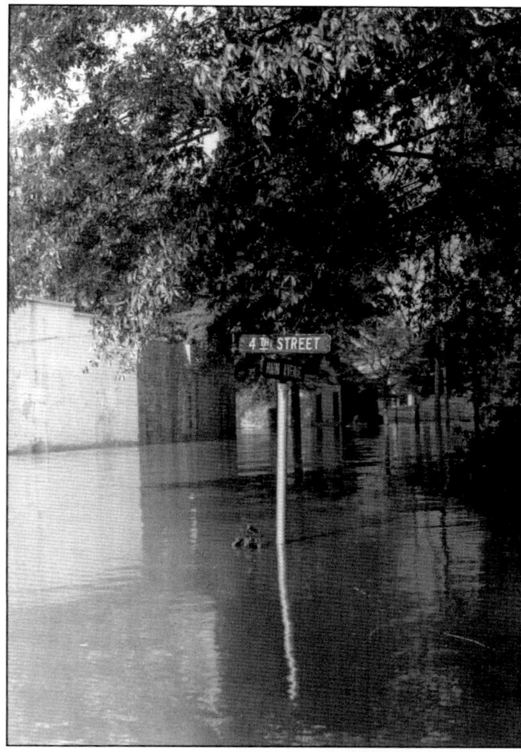

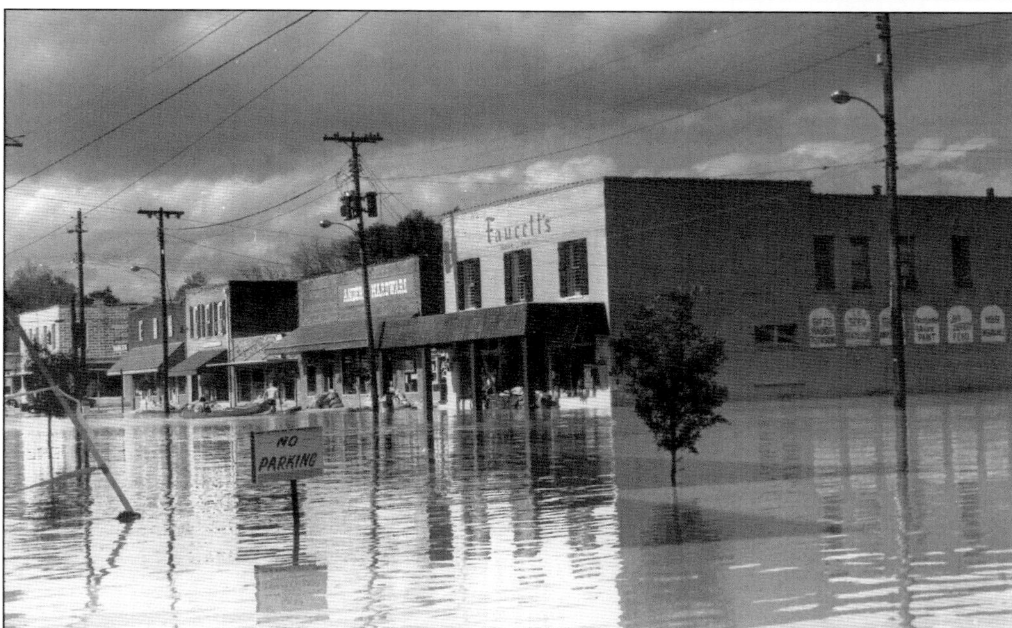

The east side of Main Avenue is seen here after the 1979 flood, with Faucett's, Anders Hardware, and Hamilton Appliance and TV Service surrounded by floodwaters. Two young men navigate the floodwaters in a canoe. All storefronts along the business district were sandbagged in an attempt to keep damage to a minimum.

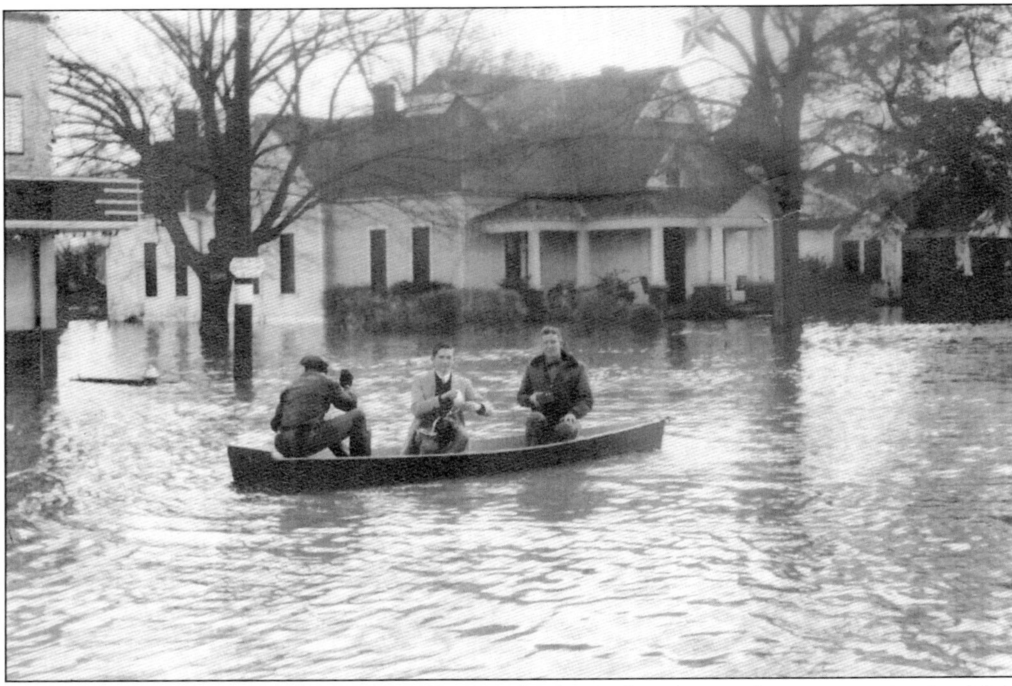

This 1950 flood photograph shows Dan Summerlin (center) and two other men navigating Main Avenue in a boat. At left is the Disney Theatre and at right is the Maxwell-Hamner House. The Black Warrior River crested above 62 feet in the 1950 flood, which was one of four major floods in four years from 1948 to 1951.

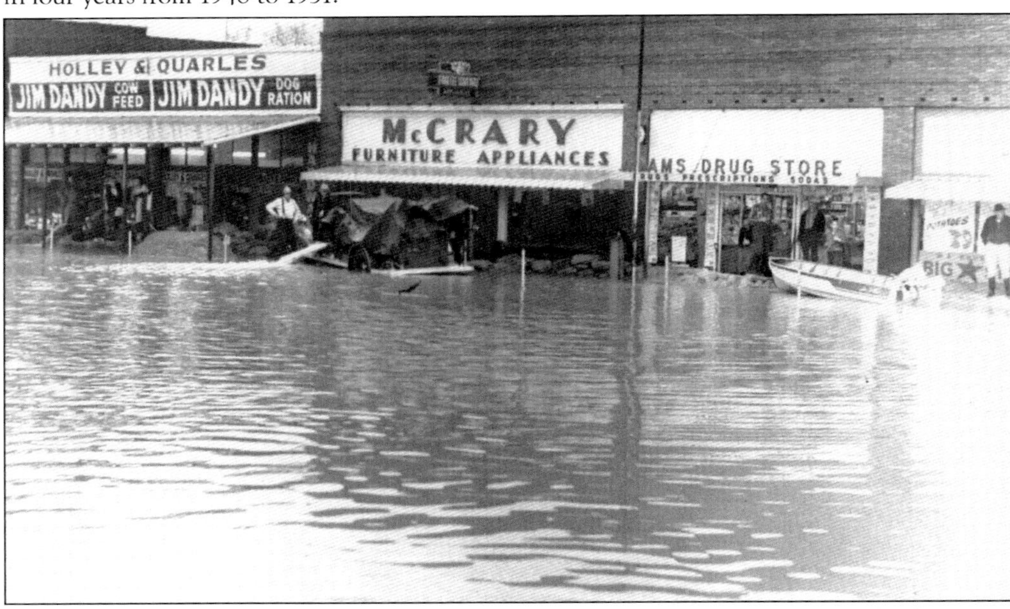

Seen in this 1961 flood photograph are Holley & Quarles, McCrary Furniture, and Adams Drug Store on the west side of Main Avenue. Fire truck pumps operated for three days during the flood to lower the water level in the stores. During that time, motorboats were the only form of transportation in the downtown area.

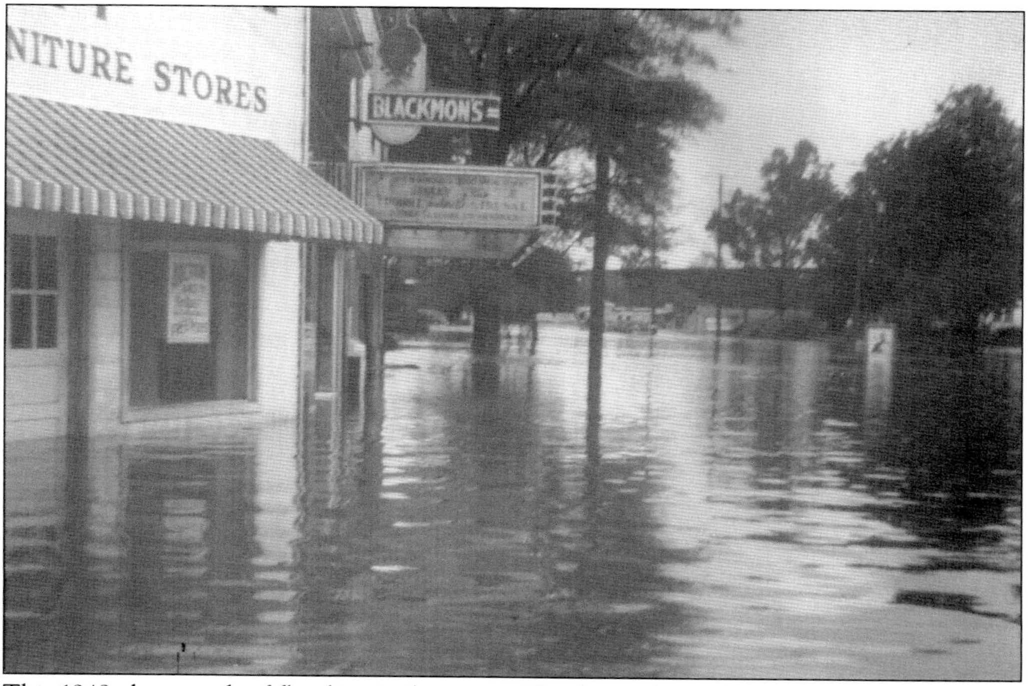

This 1948 photograph of flooding in downtown Northport shows the Furniture Auction Mart, Blackmon's Shoe Store, and the marquee for the Disney Theatre, which reads, "Double Feature on Friday and Saturday."

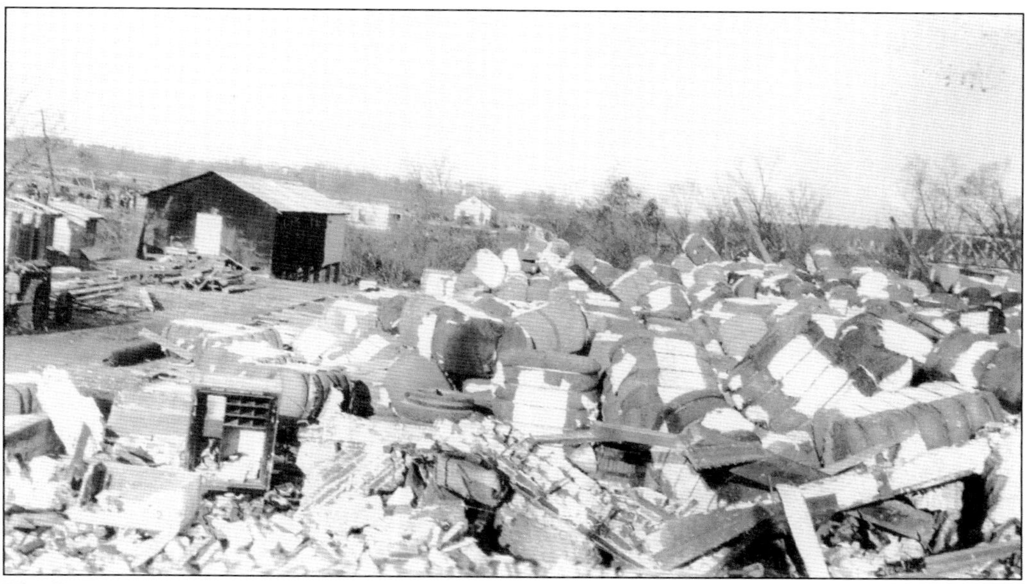

The Barnes & Norris cotton warehouses were blown away during the March 21, 1932, tornado. The gin building was also damaged, losing its roof in the tornado.

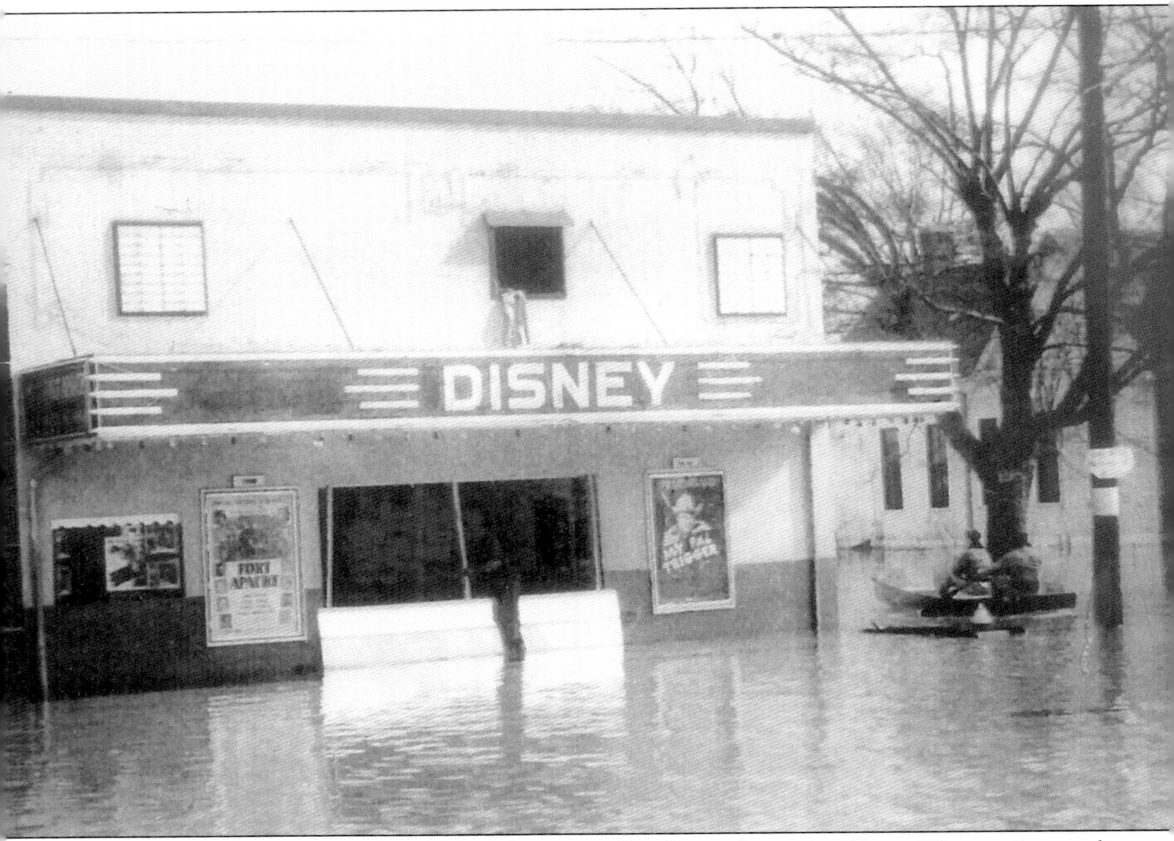

This 1948 photograph of flooding in downtown Northport shows the Disney Theatre. Located at the northwest corner of Main Avenue and Fourth Street, the 500-seat theater opened in October 1947. Prior to opening the theater, proprietor Arthur R. Disney worked in the coal business for nearly 40 years.

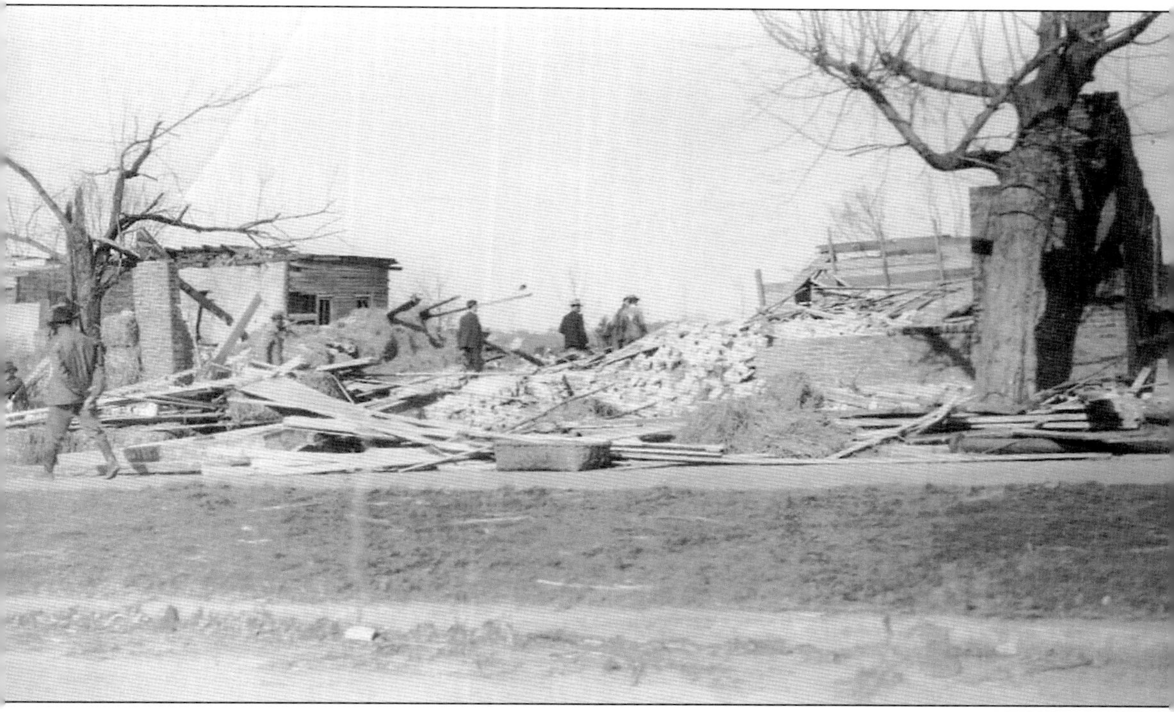

The 1932 tornado shattered brick-and-mortar structures as well as wood-framed buildings. These large oaks, planted years earlier, were torn to shreds in moments. Here, searchers look for victims and begin to assess the damage. This scene is very similar to that seen in the photographs taken in the aftermath of the April 27, 2011, tornado in Tuscaloosa.

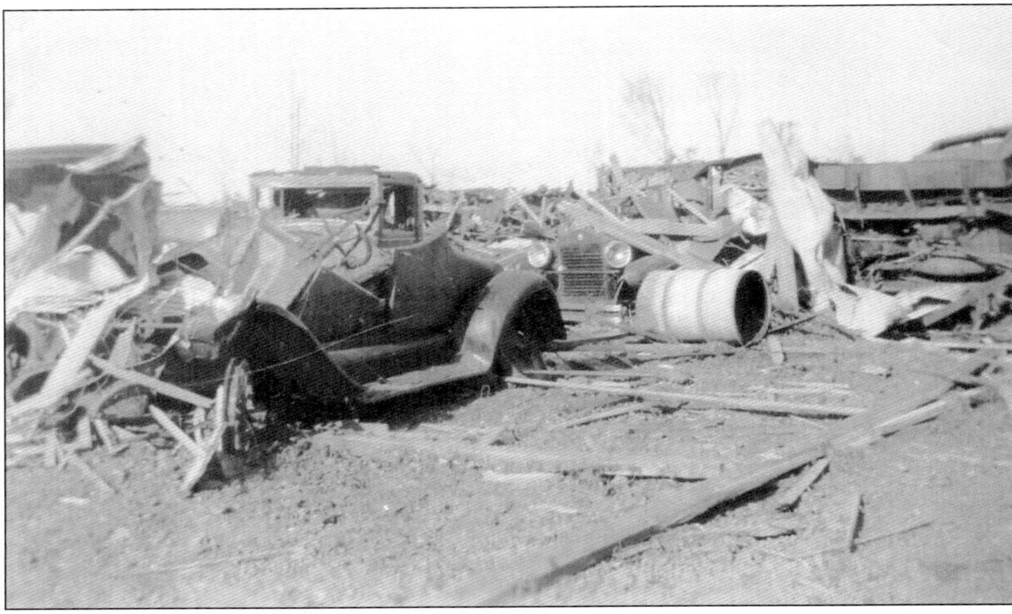

Along Northport's riverfront, anything in the path of the 1932 tornado was reduced to rubble. Note the tires on these vehicles; the powerful tornado ripped the rubber right off their rims. Also, the trees in the background were stripped bare of early foliage.

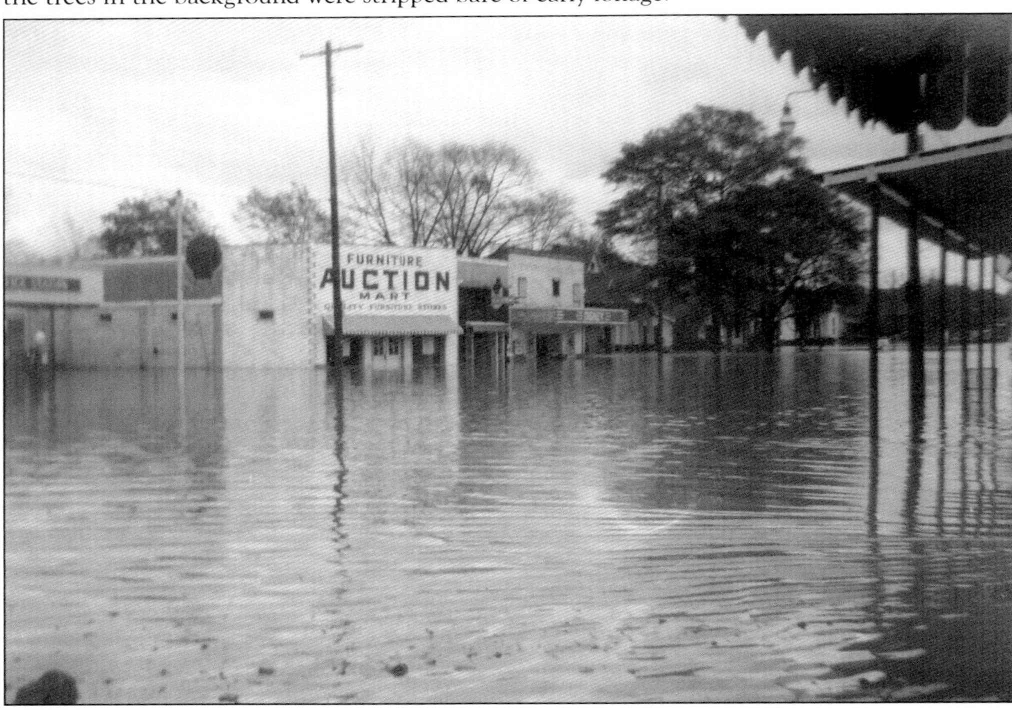

This photograph shows Faucett Bros., Strong's Service Station, and the Furniture Auction Mart—where Sue's Flowers is now located—after a flood in the winter of 1948–1949. Note the height of the water in the buildings at this stage of the flood. The Black Warrior River crested at 63.5 feet on November 29, 1948, and a few weeks later, on January 6, 1949, crested at 64.4 feet.

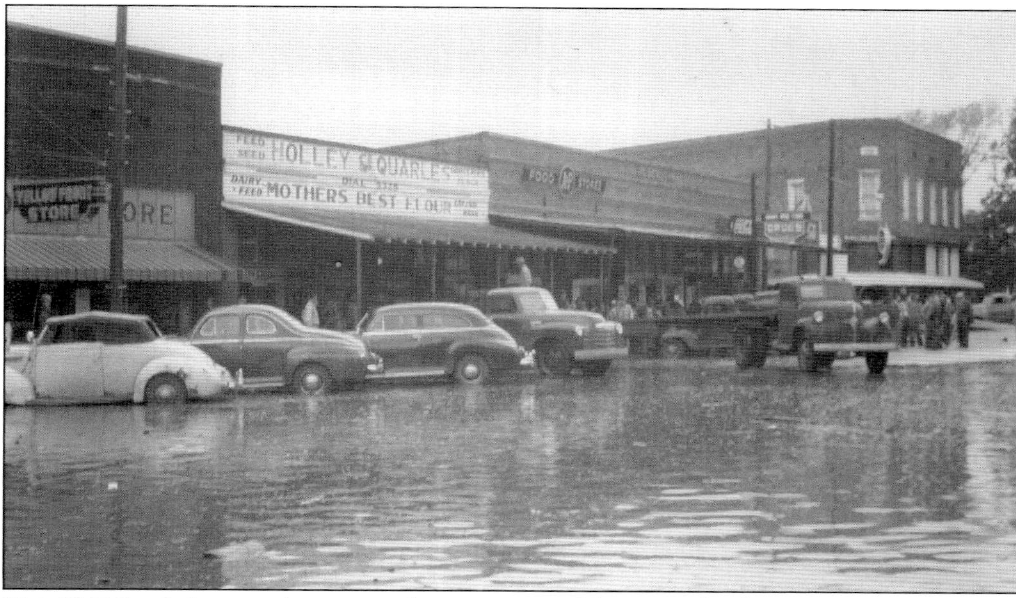

This 1948 flood photograph gives a wide perspective of the west side of Main Avenue at the time. To the left are the Yellow Front Store, Holley & Quarles, A&P Food Store, and Adams Drug Store. To the far right is the 1904 Christian Building. The floodwaters had reached the foot of Fifth Street (formerly Columbus Street).

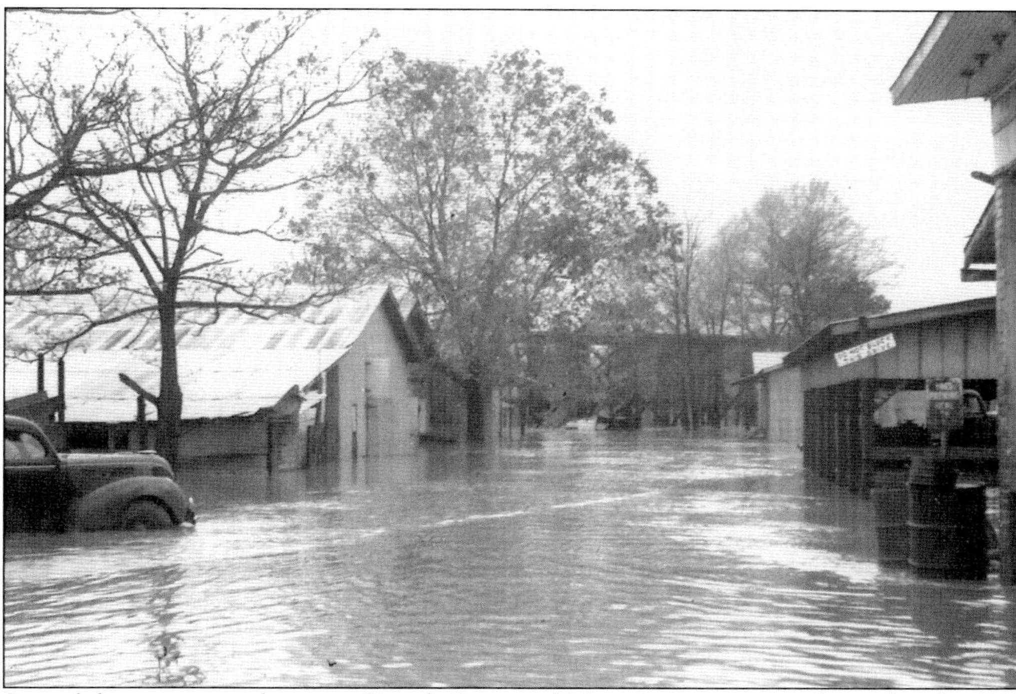

Several downtown warehouses at Fourth Street and Main Avenue are seen here after the 1948 flood. In the background is the 1897 Gulf, Mobile & Ohio train trestle. Many Northport citizens used the train trestle to walk across to Tuscaloosa during major floods when the river bridge was inaccessible.

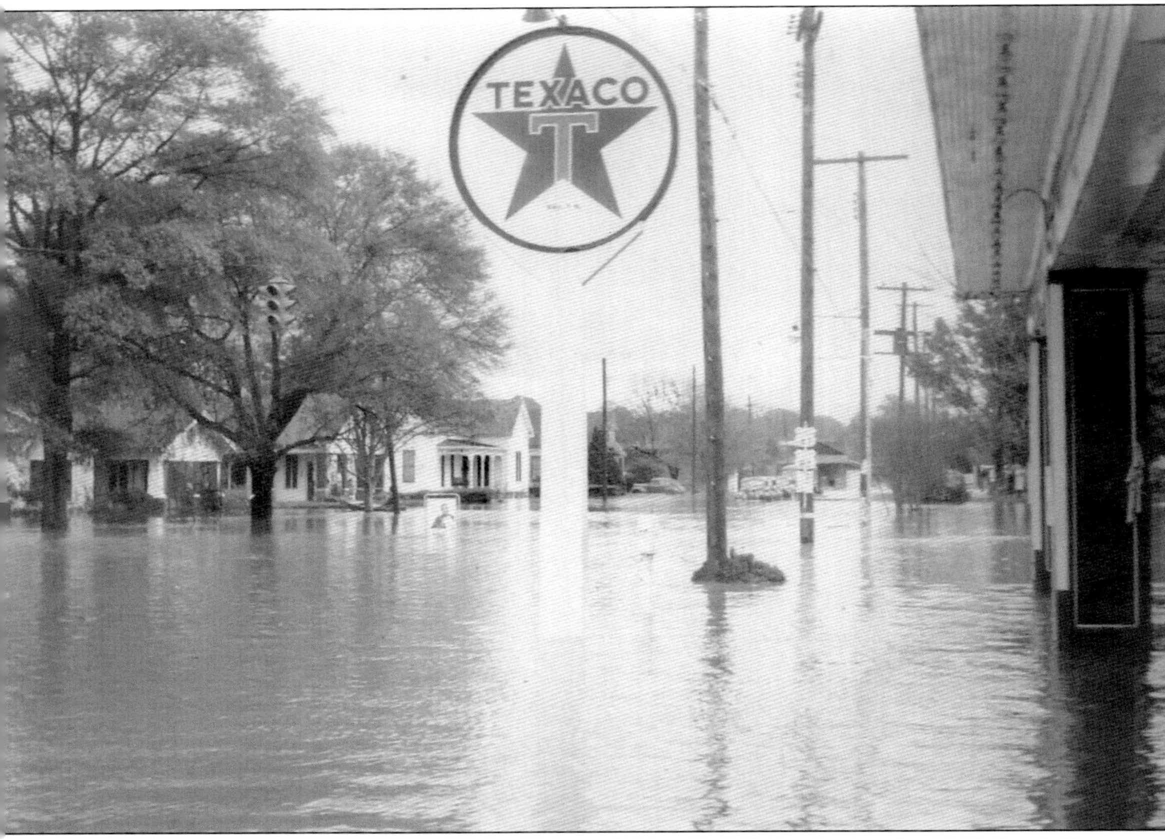

This photograph was taken in front of Maxwell Tire & Oil Co. during the 1948 flood. Down the east side of Main Avenue are several Northport residences, including the Victorian-era John Maxwell House and the Barnes & Norris Gin, at the end of Main Avenue.

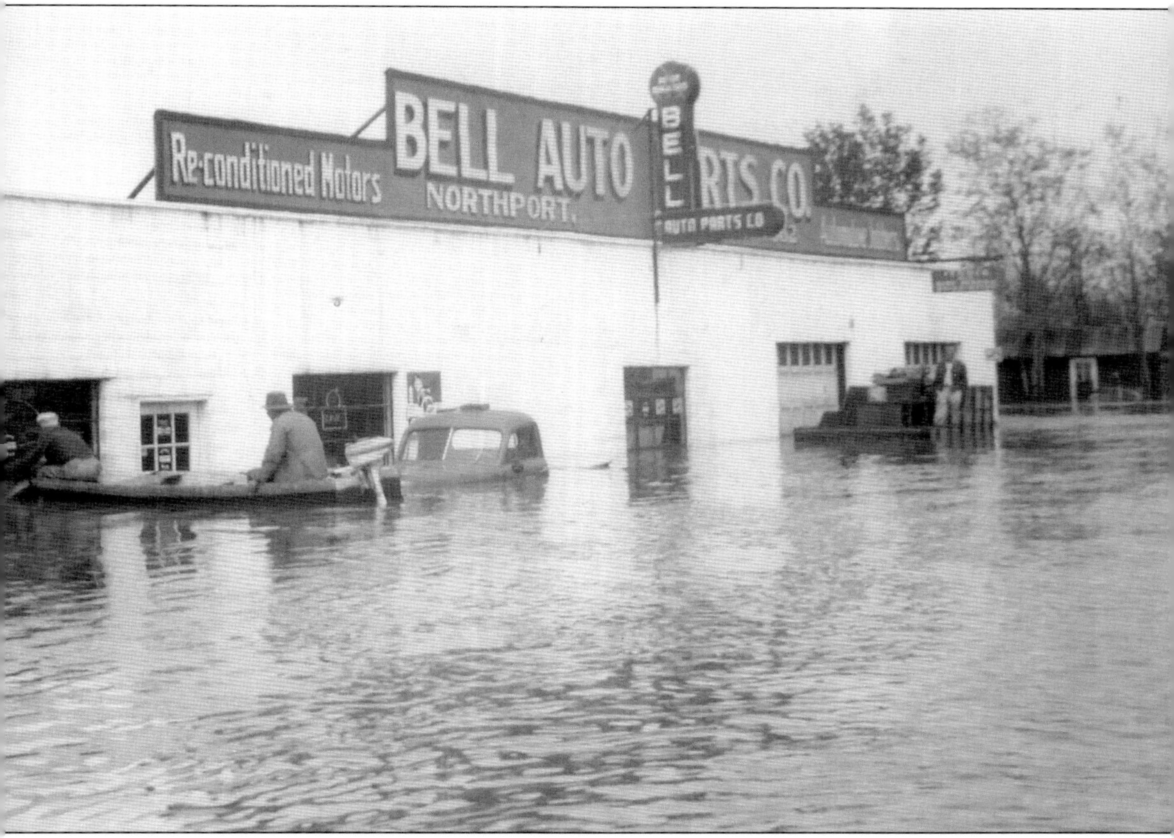

This 1948 flood photograph shows Bell Auto Parts, at 1902 First Street near the Black Warrior River, under water. The two gentlemen in the boat work to secure the building. The well-known store stood for many years at Fifth Street and the old Bridge Avenue.

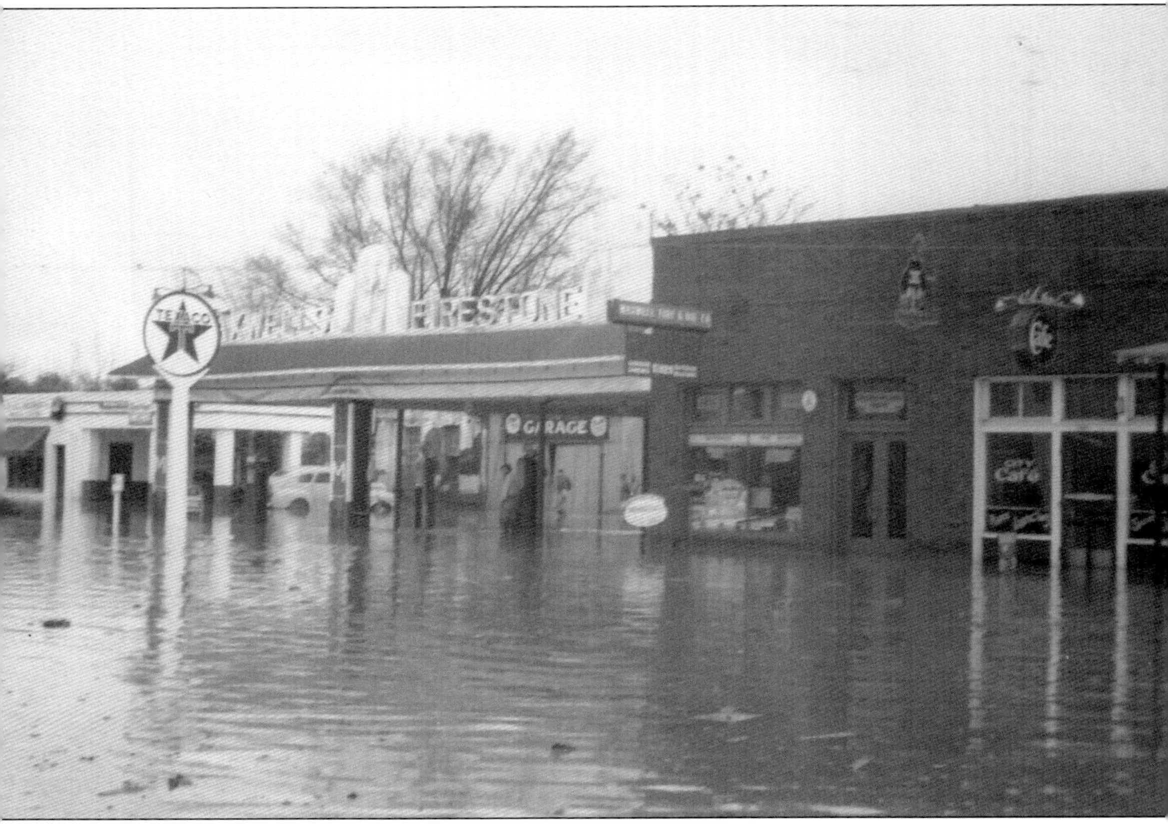

Main Avenue is seen here during the 1948 flood. The gentlemen in front of Maxwell Tire & Oil Co. are almost knee-high in river water. Thomson's service station is on the left and City Café is on the right, with a window advertisement for its plate lunches, which it continues to be famous for today.

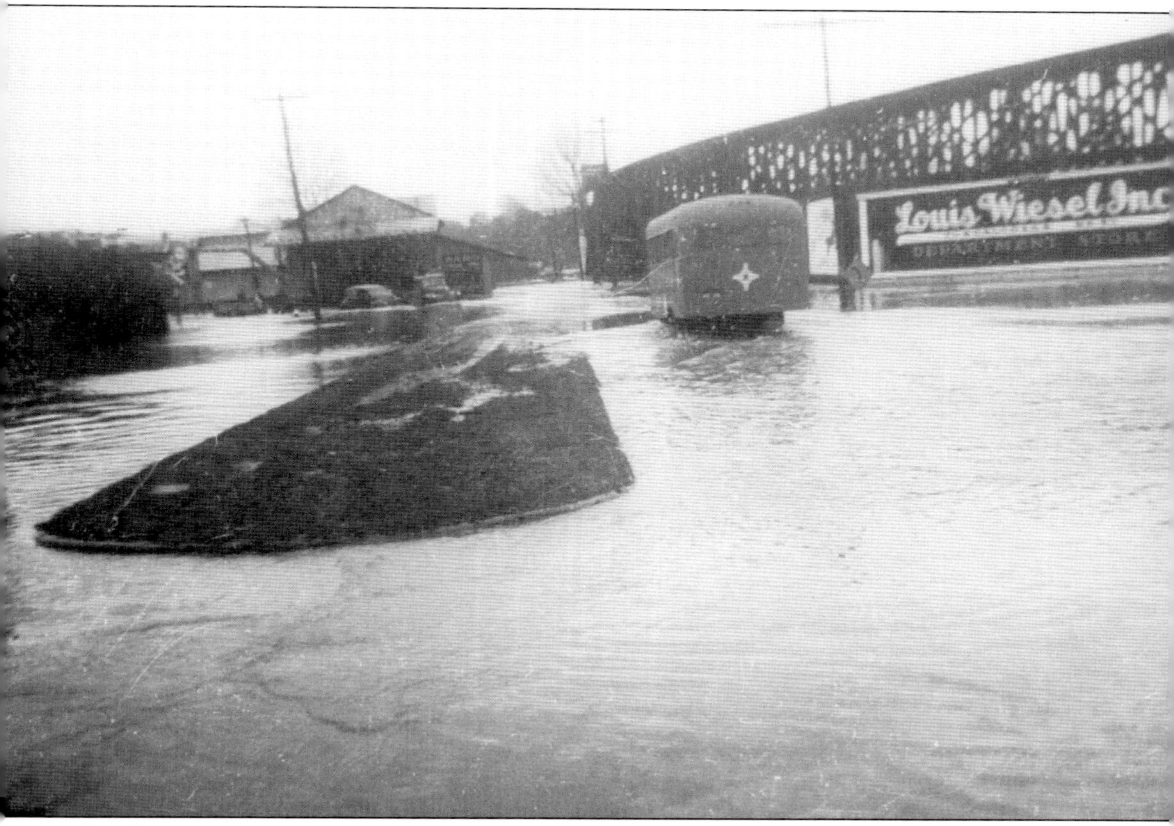

In this photograph from the 1948–1949 flood, Barnes & Norris Gin is at the end of Main Avenue. A bus navigates the floodwaters as it turns onto First Street. The grassy medians seen here were once part of the Northport street landscape.

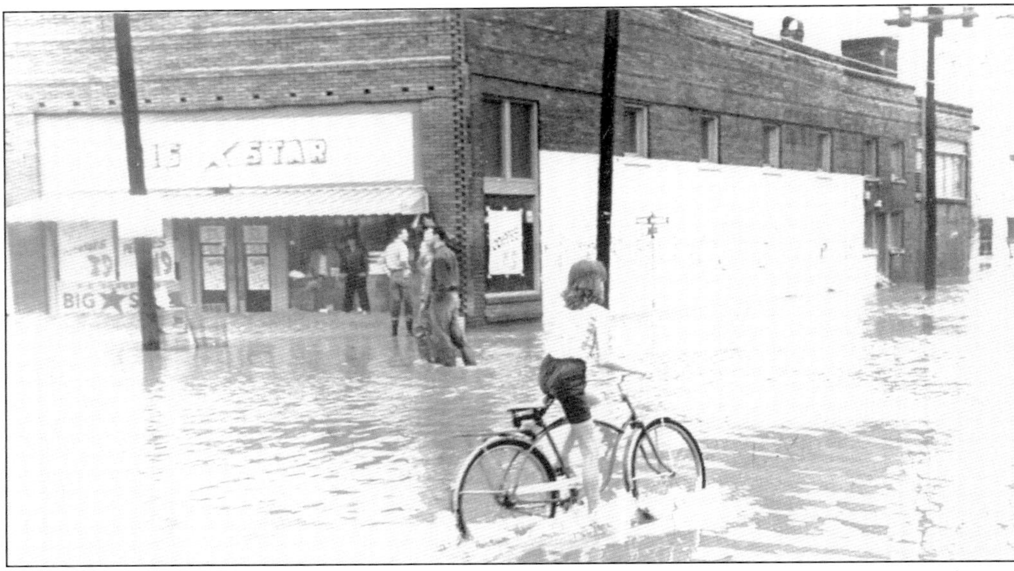

This 1961 flood photograph shows the intersection of Fifth Street and Main Avenue at the end of the second day of the flood. The Big Star grocery store on the southeast corner was operated by Paul Robertson. Signs in the windows indicate that the price of coffee was 59¢.

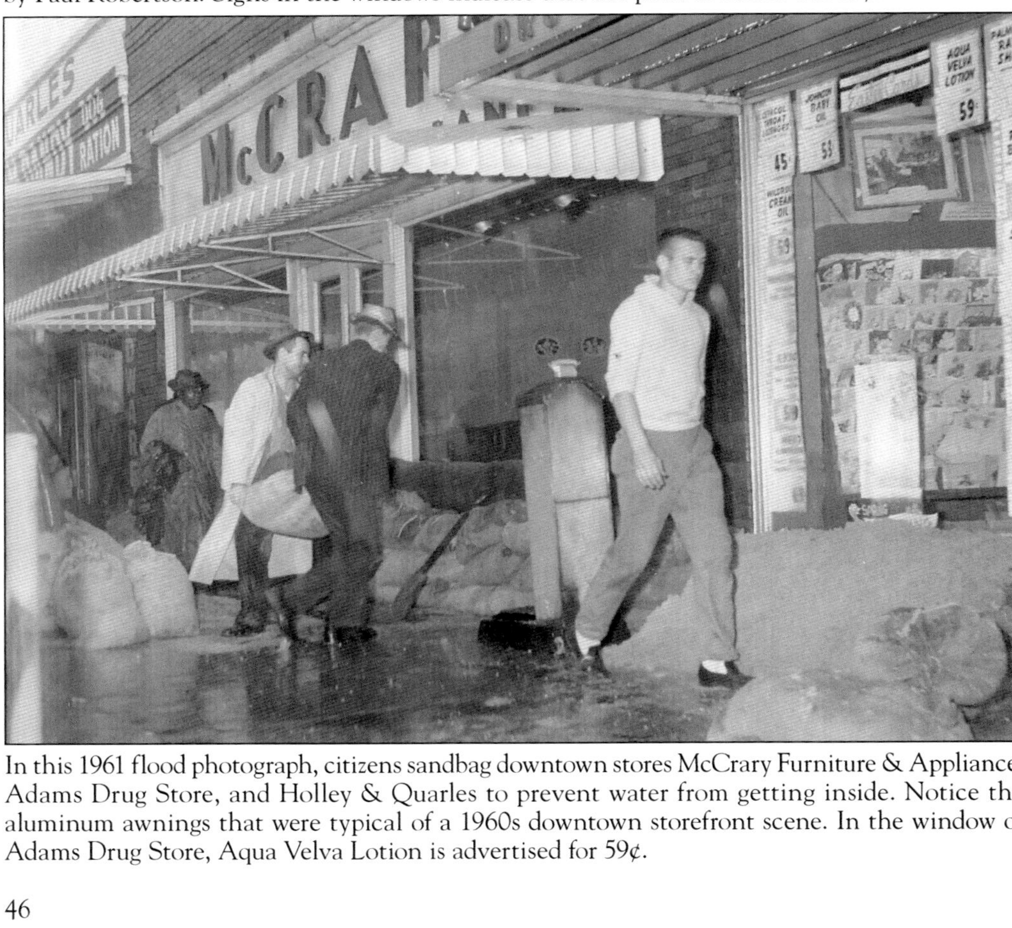

In this 1961 flood photograph, citizens sandbag downtown stores McCrary Furniture & Appliance, Adams Drug Store, and Holley & Quarles to prevent water from getting inside. Notice the aluminum awnings that were typical of a 1960s downtown storefront scene. In the window of Adams Drug Store, Aqua Velva Lotion is advertised for 59¢.

Upchurch Manufacturing was located on south Main Avenue near First Street. The business, owned by Charles B. Upchurch, manufactured interior home items, upholstery, and mattresses. This 1948 flood photograph shows the building under water while two gentlemen work to secure items from further damage.

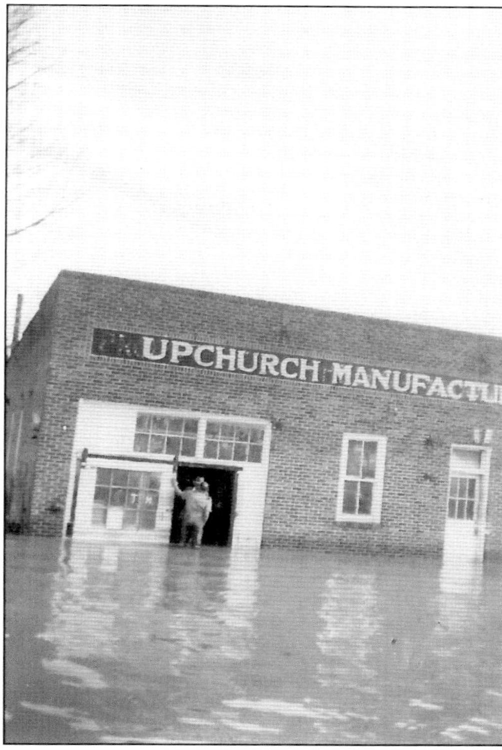

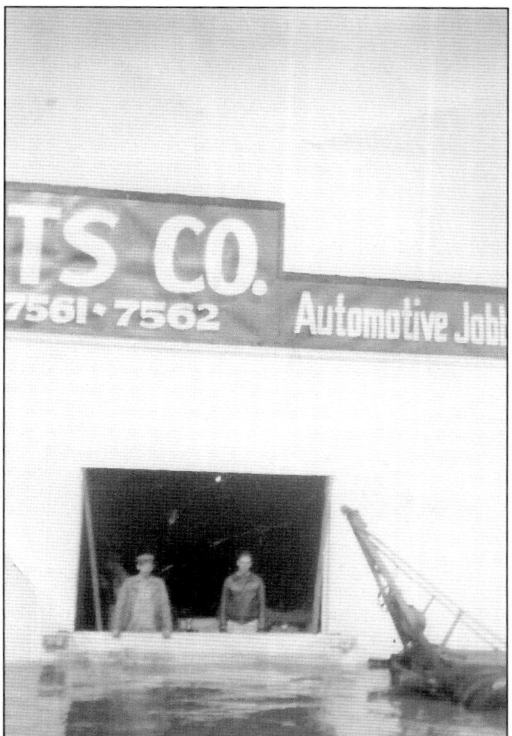

The 1948 flood affected Bell Auto Parts, at 1902 First Street near the river. The business was a partnership owned by Frances Bell, Bob Delbridge, and George Bell. It was started in 1939 and remained on First Street until 1988, when the property was sold and the business was moved to Fifth Street and Bridge Avenue. This photograph was taken by Mary L. Bodiford.

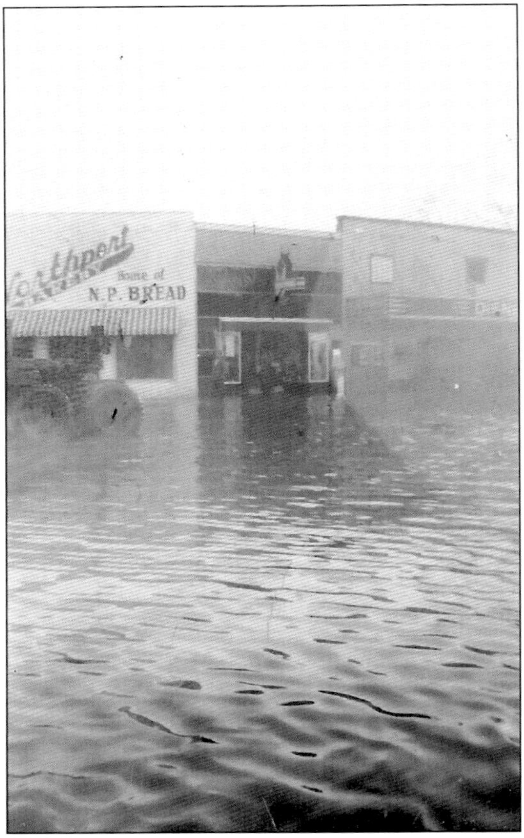

This photograph from the 1948 flood shows the old Northport Bakery, located where Sue's Flowers is today. Next door was Blackmon's Shoe Store and the Disney Theatre. Leon Blackmon opened his store in the mid-1940s. It was known as the "Red Goose Shoe Store" for its large sign advertising Red Goose shoes.

Seen here riding on a tractor in the streets of downtown Northport during the 1948 flood are, from left to right, (first row) Billy Pullen, Bibb Watkins, Addison Thompson, and Gerald Wiggins; (second row) Billy Dickinson, Sam Palmer, and Billy Deal Howell. Mary L. Bodiford took this and other photographs of the flood while traveling to work on a city bus.

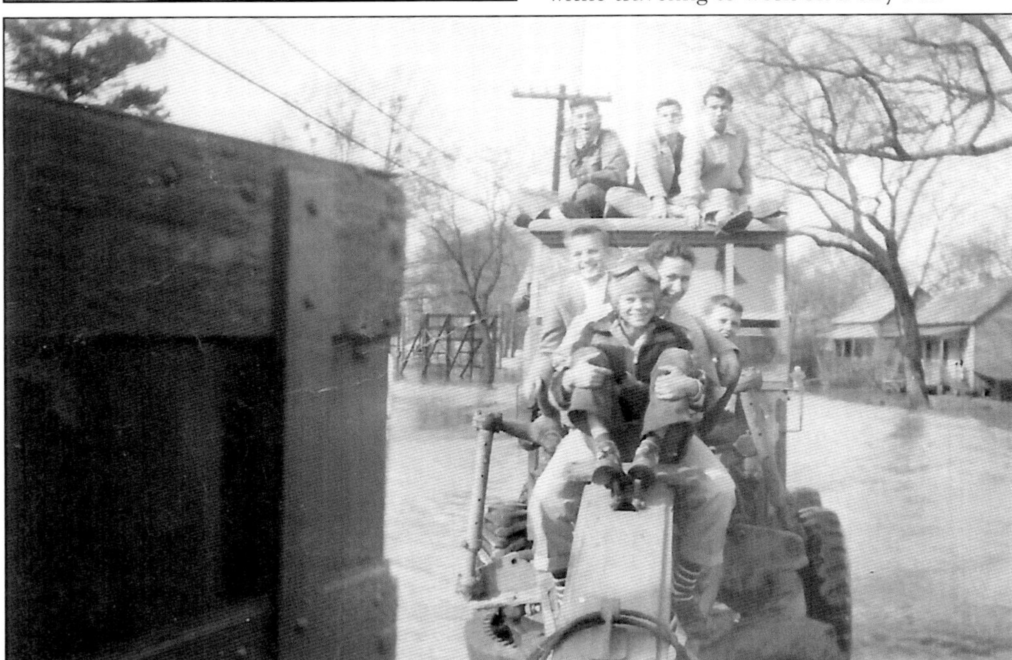

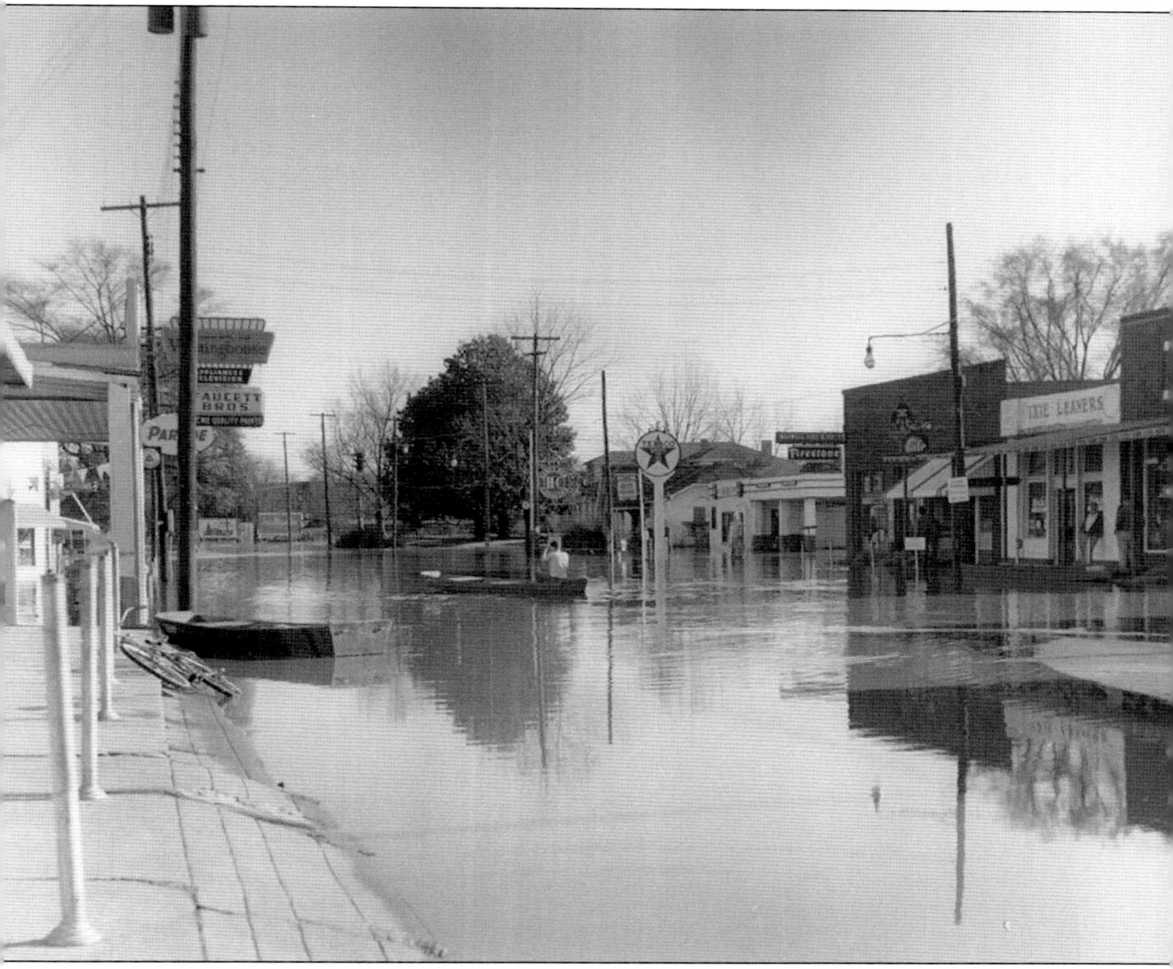

In this 1949 flood scene, a gentleman navigates a boat south along Main Avenue. The stores seen here include Faucett Bros., Maxwell Tire & Oil Co., Dixie Cleaners, and Adams Drug Store. In the background are the Sinclair Station, the home of Northport mayor James Anders, and the Maxwell-Peters home.

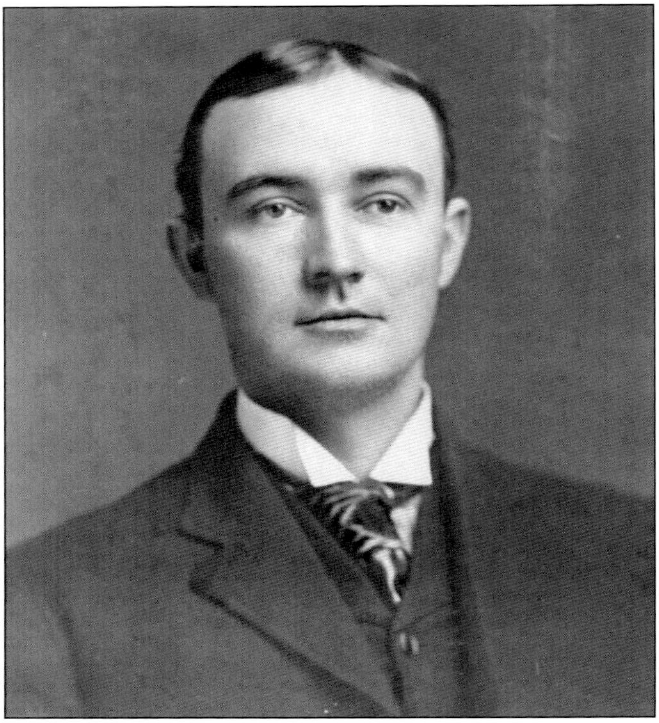

H.G. Shepherd served on the city council from 1920 to 1924. He was elected mayor of Northport in 1932, serving until his unexpected death in 1938. Shepherd lived on Twenty-fifth Avenue in Northport in a two-story frame house built in the 1890s. The 1932 tornado severely damaged the upper level of the house, and when the building was repaired, the top floor was left off.

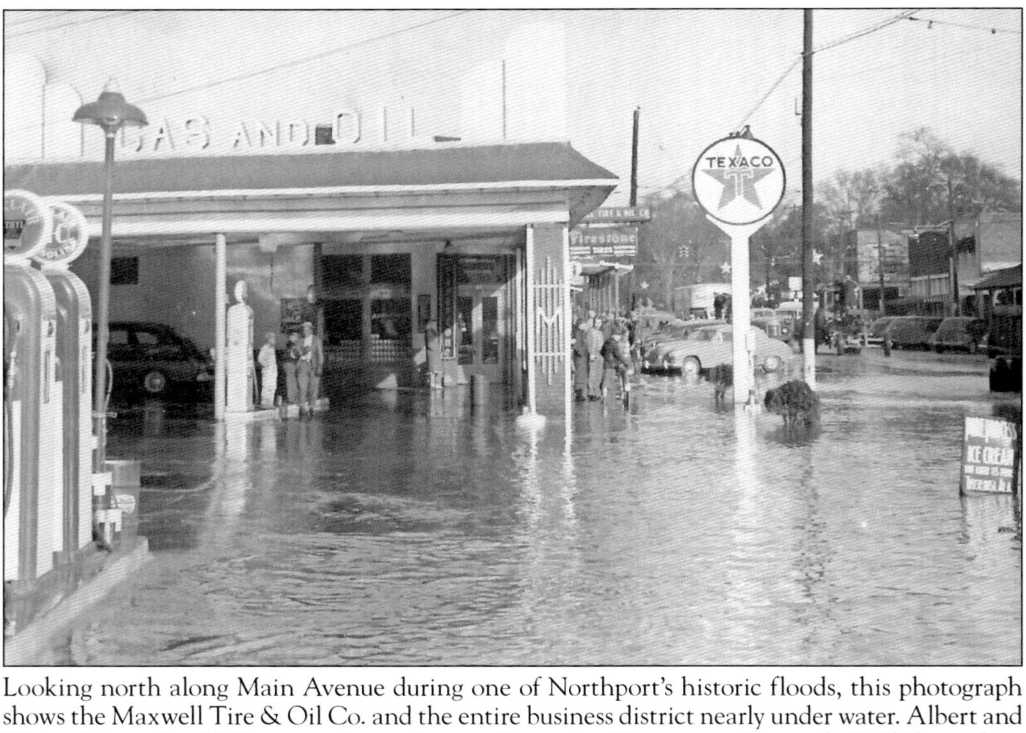

Looking north along Main Avenue during one of Northport's historic floods, this photograph shows the Maxwell Tire & Oil Co. and the entire business district nearly under water. Albert and Walter Maxwell sold Firestone tires, tubes, and batteries; Texaco gasoline and oil; Philco radios, refrigerators, and televisions; and Sherwin-Williams paints.

Three

Live, Work, and Enjoy

When it comes to being a successful community, Northport offers the three necessary qualities to attract and keep a growing population: live, work, and enjoy. Northport has exceptional healthcare facilities, outstanding schools, growing and diverse neighborhoods, an abundance of recreation and park facilities, welcoming houses of worship, great restaurants, and a thriving arts community.

The following essay, "Why I Love Northport" by fifth-grader Rayleigh Stripling, was presented at the May 2009 Friends of Historic Northport Annual Heritage Homecoming Luncheon.

> Northport, Alabama is a beautiful, All-American city full of rich heritage. Northport is the home of many extravagant buildings and people. I love Northport because of the fun and great places, the many occasions enjoyed by citizens of Northport, and also because of the breathtaking scenery.
>
> The first reason why I love Northport is because of the fun and great places the city has to offer. An example of fun and great places is the restaurants. My favorites are Taco Casa and City Café. People have a lot of fun at the City Café. For instance, my dad and pop used to take me to City Café all the time when I was little. My final example is the parks, such as Kentuck Baseball Park and the Northport Community Center Park where people enjoy their weekends.
>
> The second reason why I love Northport is because of the special occasions often enjoyed by many of the Northport citizens. An example of this is Dickens Downtown where many people gather in downtown Northport to celebrate Christmas as their great-great-grandparents did. Another example is the Kentuck Festival where people come from all over to see the works of folk artists. My final example is the Christmas on the River, an occasion where boats are decorated and float down the river for everyone to see.
>
> The third and last reason why I love Northport is because of the unfathomable scenery. Most of the beautiful scenery comes from the waterways, like Lake Tuscaloosa, Lake Lurleen, and the Black Warrior River. Historic Downtown Northport also has a lot of amazing scenery, like the brick streets that clack under your feet and the gorgeous trees that sway in the breeze. Another example of scenery in Northport is Hidden Meadows Golf Course. It has many breathtaking trees and ponds on the golf course.
>
> Northport is a delightful, All-American city filled with rich heritage. This city is home to many dignified buildings and people. I love Northport because of the many fun and great places, the occasions enjoyed by many people, and all of the amazing scenery it has to offer.

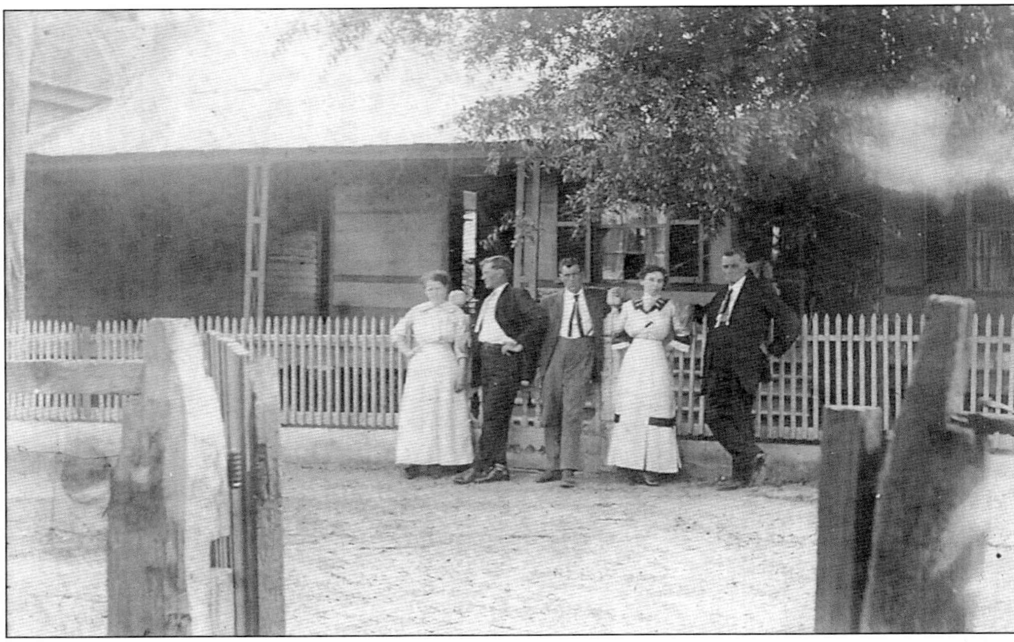

Members of the Moore family gather in front of the home of John Moore in 1902. From left to right, they are Beulah Ramsey, J. Mitt Ramsey, Will Moore, Estelle Moore, and John Samuel Moore. In 1909, there was a millinery shop in one room of this house, on Columbus Street (now Fifth Street) near the train trestle.

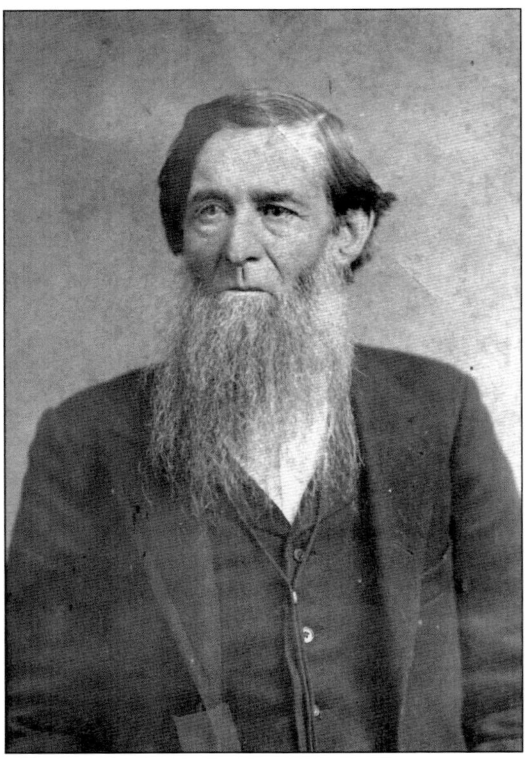

Reuben Dodson was the pastor of Northport Baptist Church from 1857 to 1860. Dodson, a South Carolina native, married Elizabeth Anders in 1827. He and Otis Dyer surveyed the downtown Northport area and sold it off in lots. The Dodson and Dyer Survey is still used today.

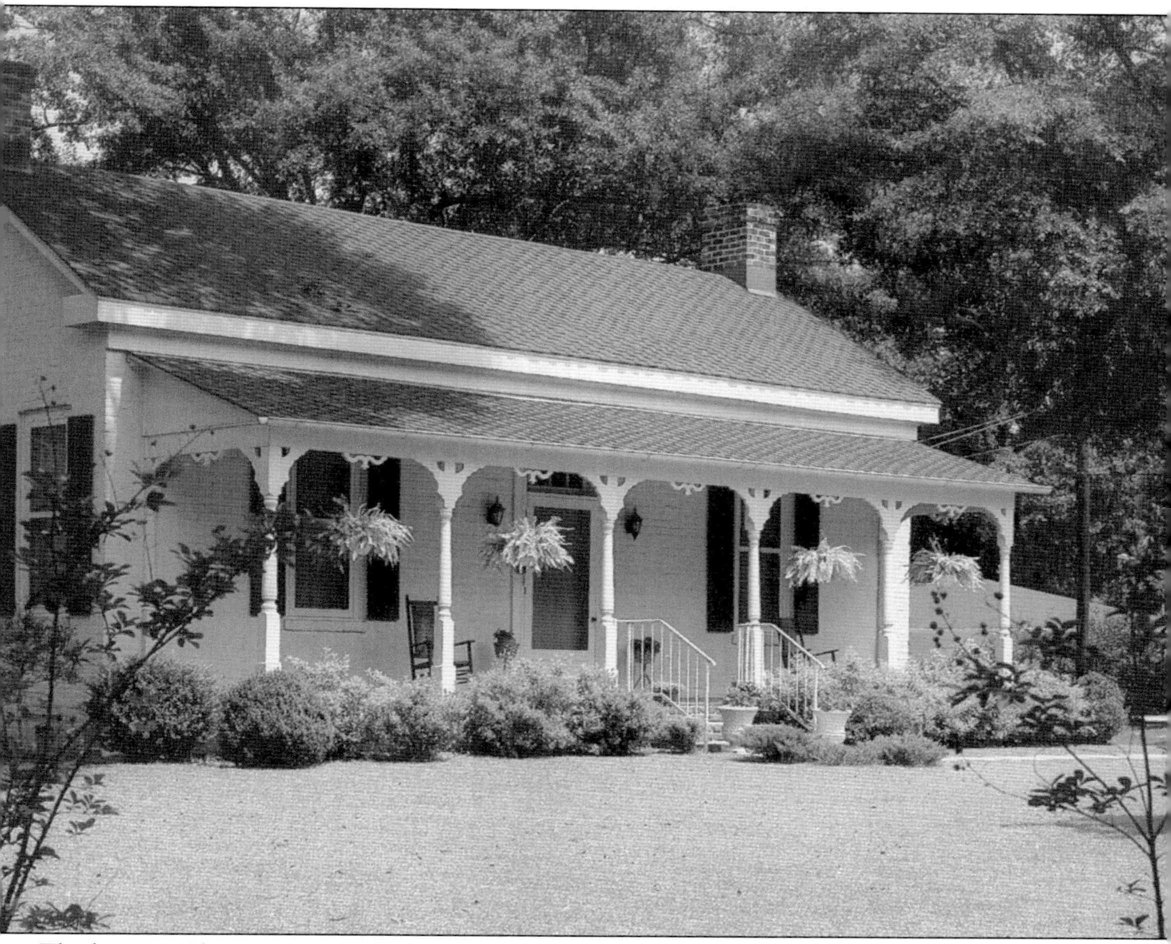

The historic Clements House, built in the mid-1830s, is located at 1802 Twentieth Avenue. In 1984, W.W. "Foots" Clements completely restored the home before giving it away for use as a historical site. A native Northporter, Clements became president, chief executive officer, and chairman of the board of the Dr. Pepper Bottling Company.

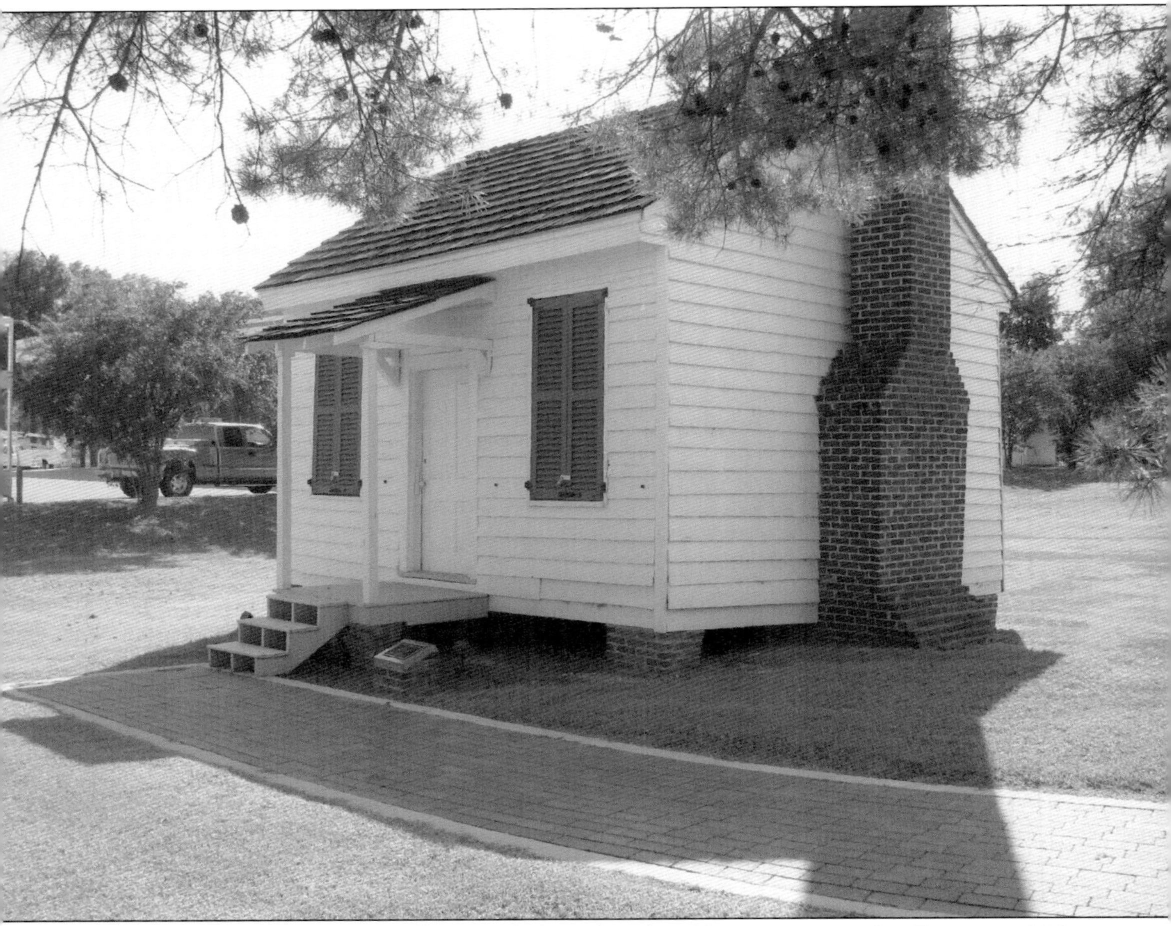

The Umbria one-room schoolhouse is believed to the oldest surviving schoolhouse in Alabama. It was built in 1829 for the Pickens family, who owned the Umbria Plantation in Sawyerville, Alabama. Jack Warner moved the structure to North River in 1973 and it was donated to Friends of Historic Northport in 2006 by the Tuscaloosa County Preservation Society. It was then placed in the Northport Community Center Park as a historical attraction.

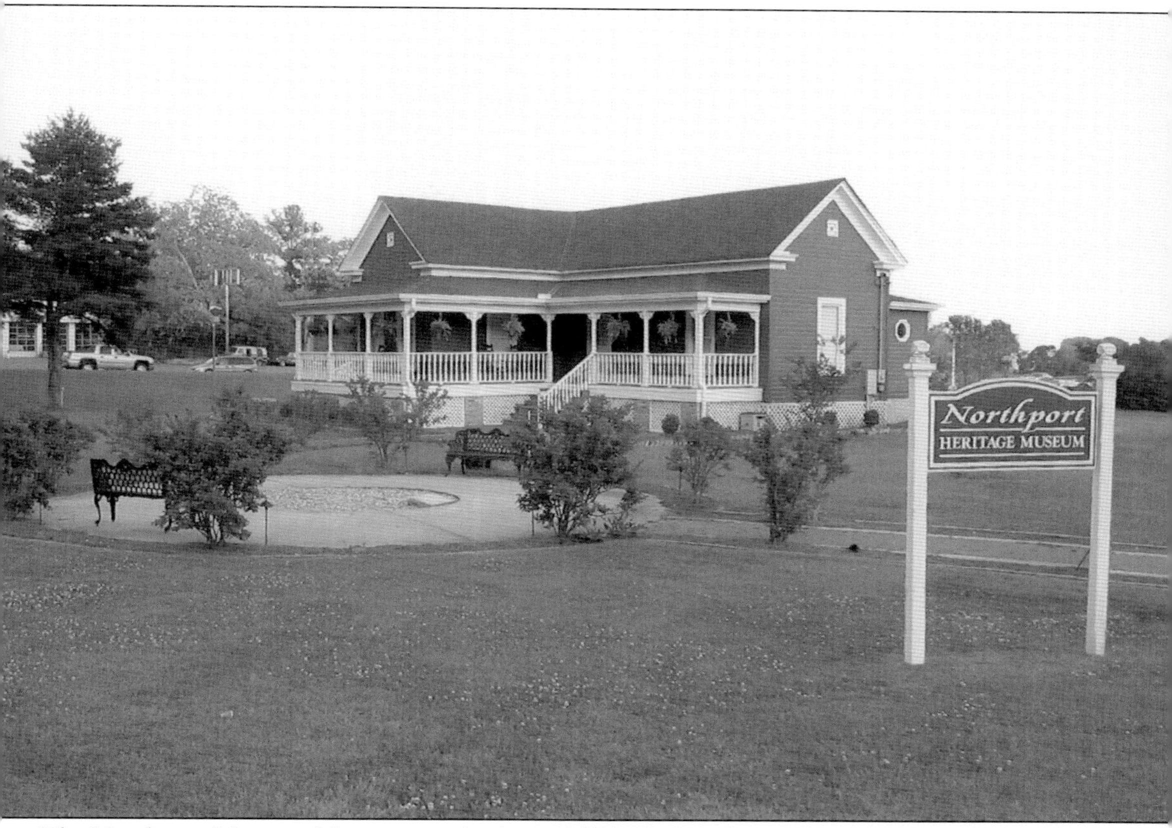

The Northport Heritage Museum was built in 1907 by Josh Palmer. The Palmer family home stood on Tanyard Street (now Tenth Street) until 1998, when it was relocated to the Northport Community Center Park as a museum. Members of the community donated money to restore the structure. It was dedicated on October 28, 2001.

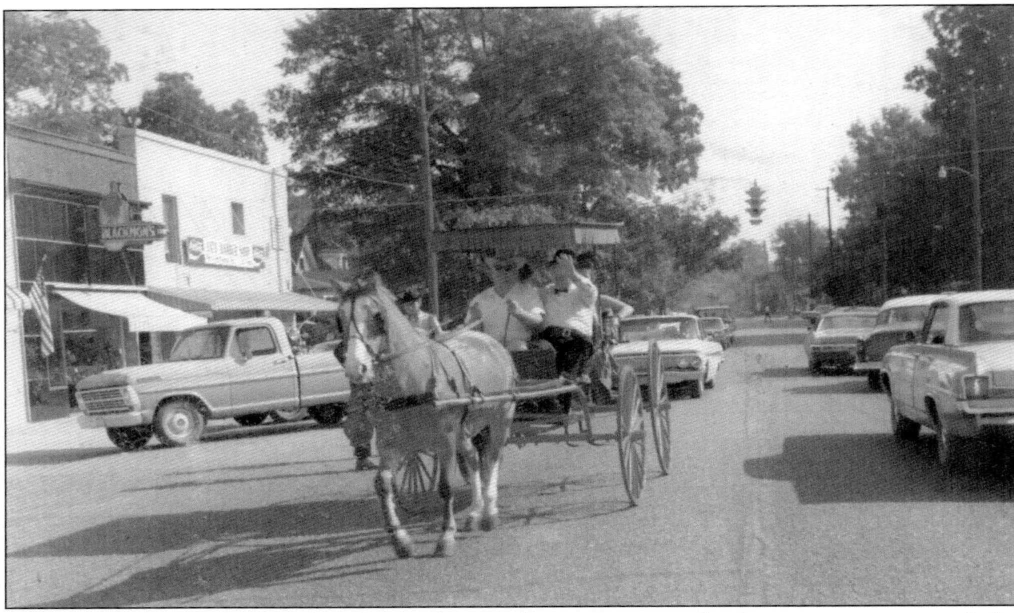

This 1971 photograph was taken during the Northport Centennial, when reenactments included horse-and-buggies like this one. To the left are Blackmon's Shoe Store and City Barber Shop, in the old Disney Theatre. The traffic signal seen here was in use at Fourth Street and Main Avenue for many years. Notice the large oaks lining the street.

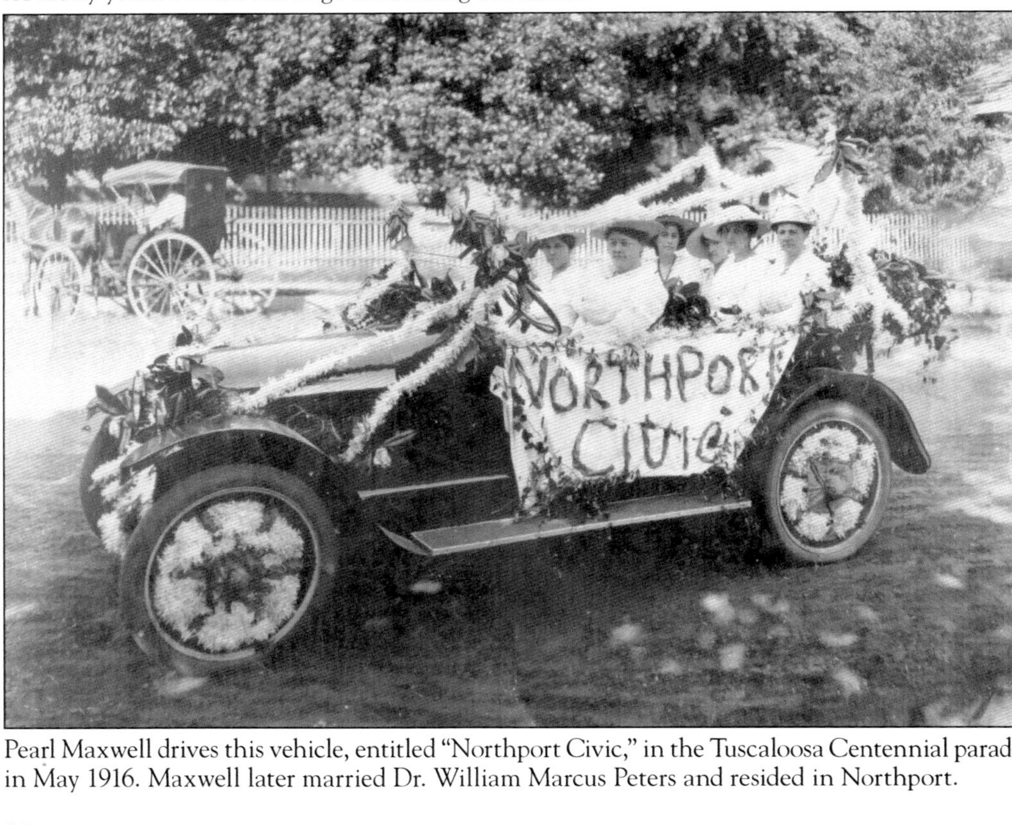

Pearl Maxwell drives this vehicle, entitled "Northport Civic," in the Tuscaloosa Centennial parade in May 1916. Maxwell later married Dr. William Marcus Peters and resided in Northport.

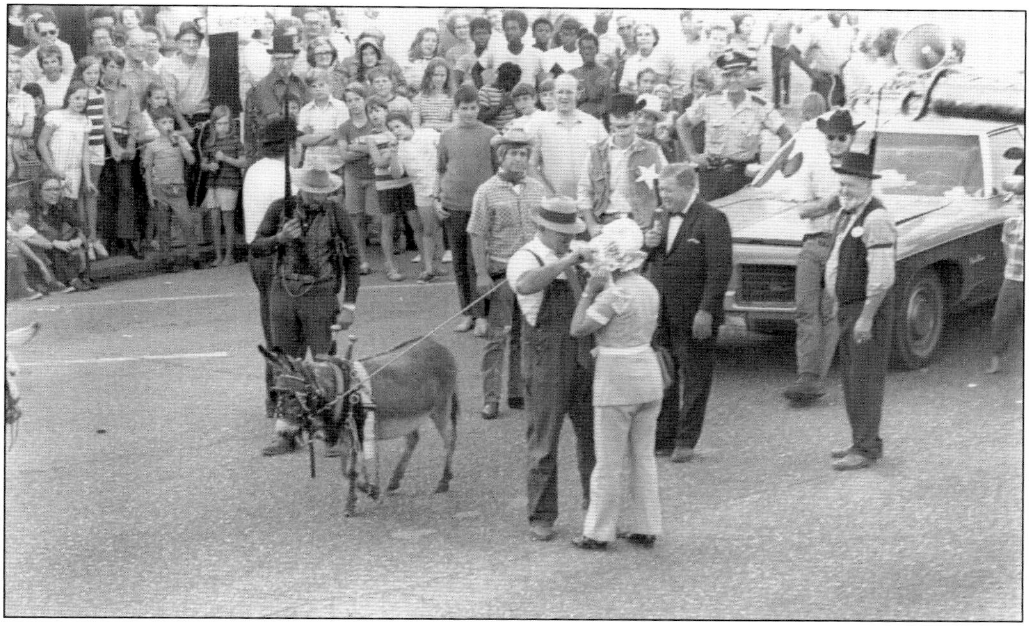

During the week of October 10–16, 1971, Northport celebrated its 100th anniversary as an incorporated town. Among the events held during the Centennial was a street play entitled *Shootout on Main Street*. As a result of this successful event, a Northport Fall Festival featuring handicrafts and artwork was planned for the following year. This fall festival eventually became the Kentuck Festival of the Arts.

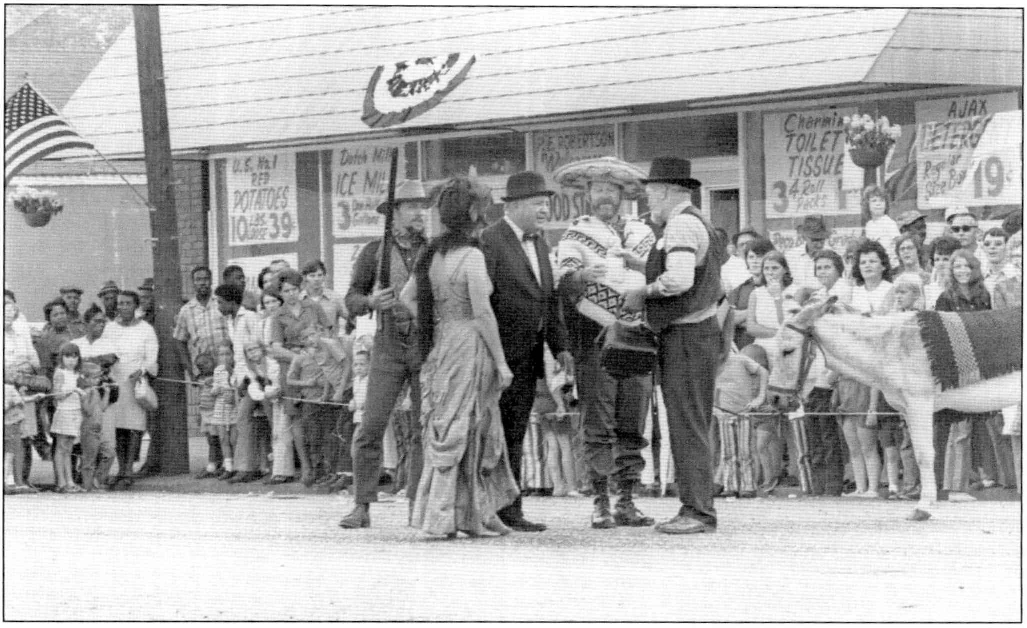

Northport's 1971 Centennial celebration included this scene from the street play *Shootout on Main Street*. In the center of Main Avenue are the "bandits" of the play. Shown in the background are Northport citizens watching the play from the sidewalk in front of P.E. Robertson's grocery store. An advertisement in the store window reads, "U.S. No. 1 Red Potatoes 10 lbs. loose 39¢."

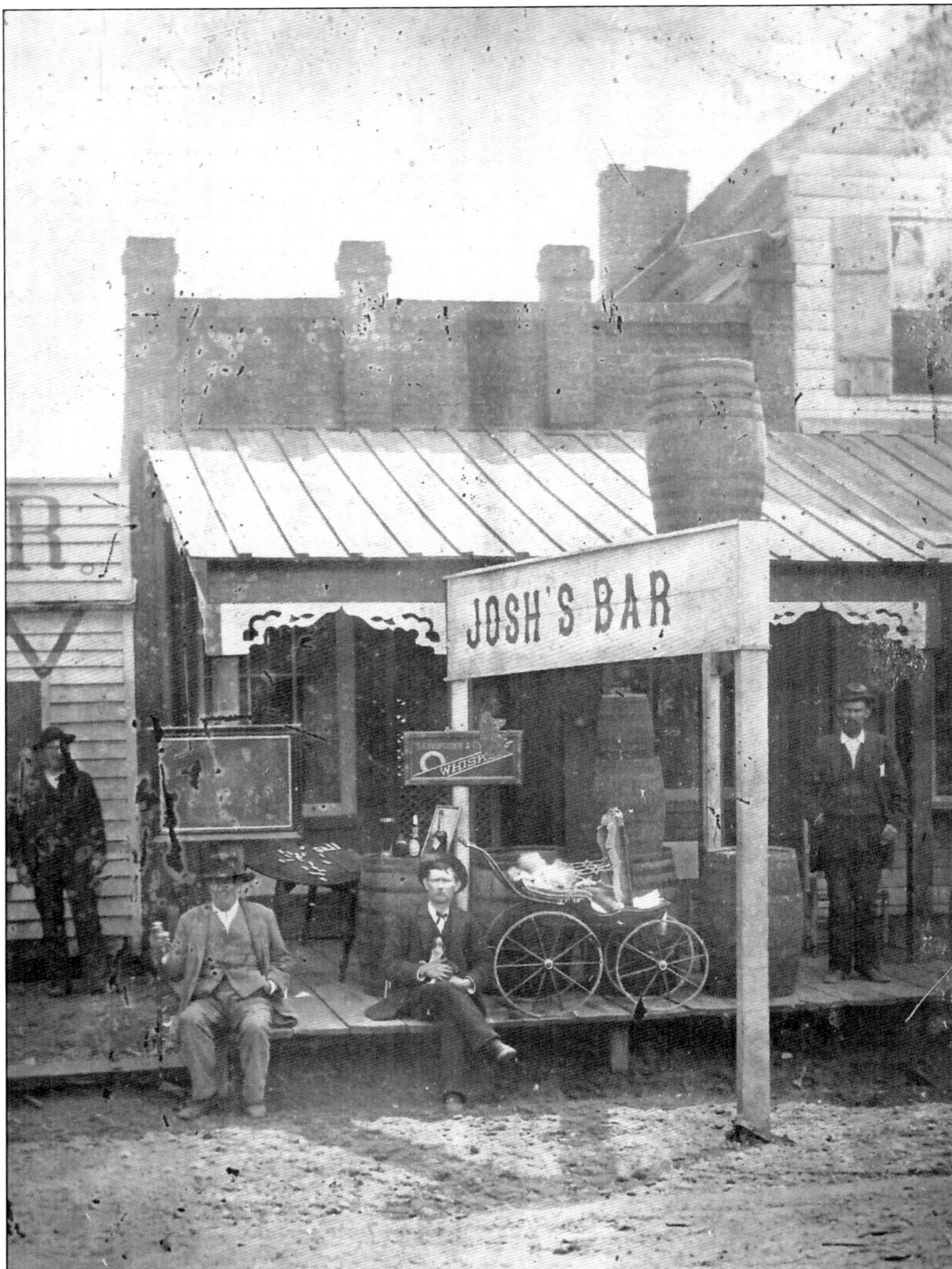

This photograph of Josh's Bar in downtown Northport is dated January 2, 1885. The establishment is named for its owner, Ezekial Joshua Palmer. Born in 1854, he was married to Mary Lee Simpson and is seen here next to a baby carriage holding Richard Terrell Palmer. The bar was located on the east side of Main Avenue in the middle of the block, where the Rice Bros. office is today. Note the large wooden barrels on the sidewalk and on the sign.

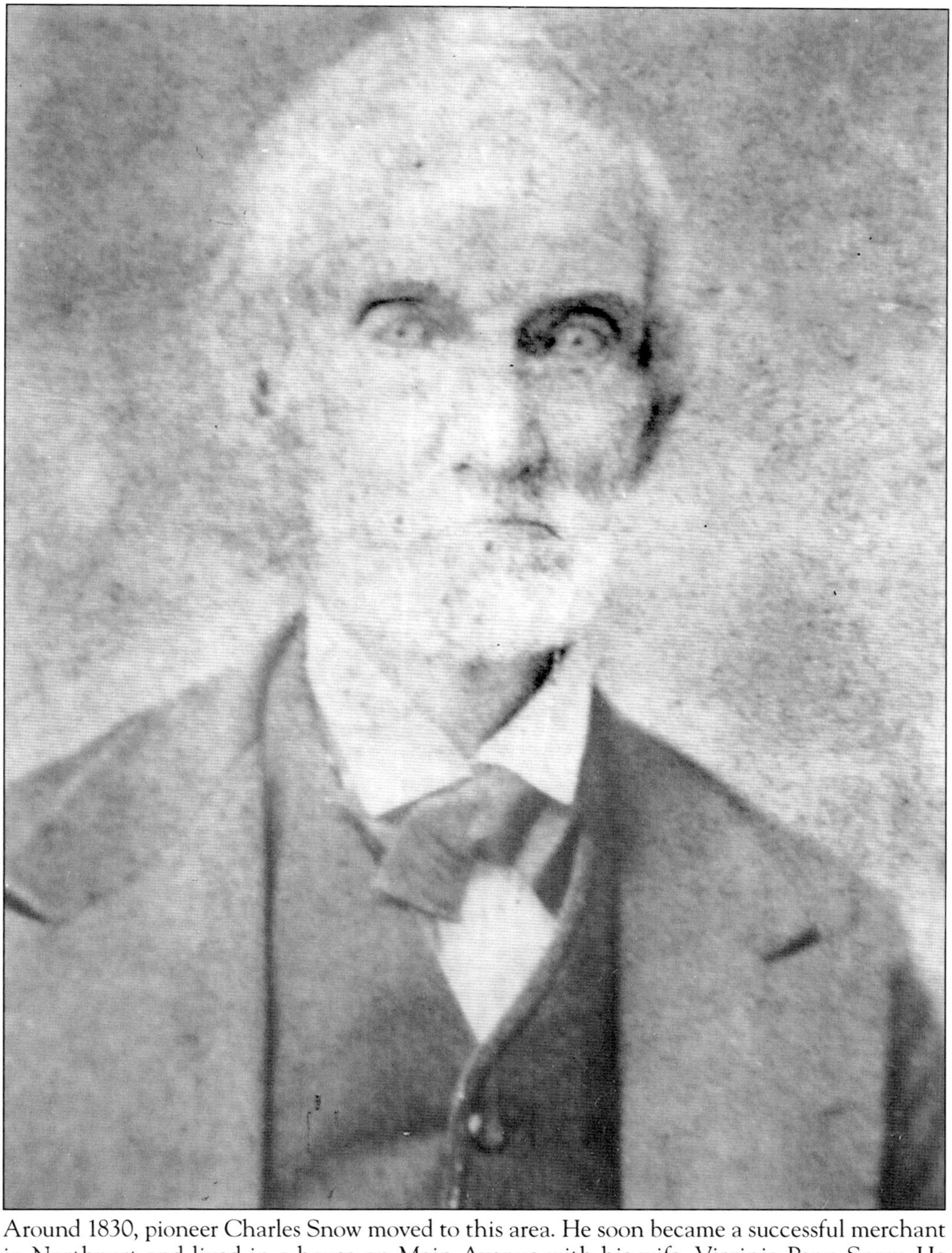

Around 1830, pioneer Charles Snow moved to this area. He soon became a successful merchant in Northport and lived in a house on Main Avenue with his wife, Virginia Penn Snow. He served as Northport's first postmaster and owned a business along the river, built of rough lumber, which had a sign above the door reading, "Charles Snow & Company, Cash Store." Snow was a very good businessman who prospered even during the bank panics of the 1830s. Later in life, he moved to the old Job Going Academy north of what is now called Snow's Mill Creek, and named it Snow Place.

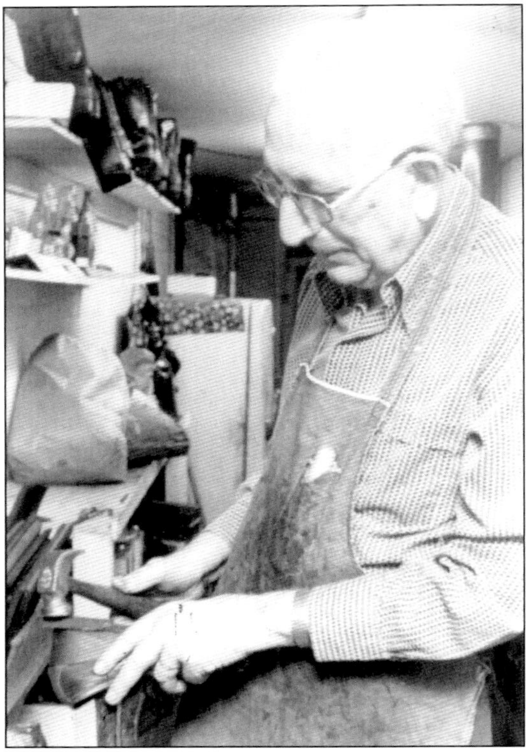

This 1960s photograph shows A.W. "Steve" Stephens in the Northport shoe repair shop. Stephens worked at Brown's Dollar Store in Tuscaloosa for several years before opening his Northport operation, first inside Anders Hardware, and later in an enclosed firebreak between Anders Hardware and Faucett's on the east side of Main Avenue.

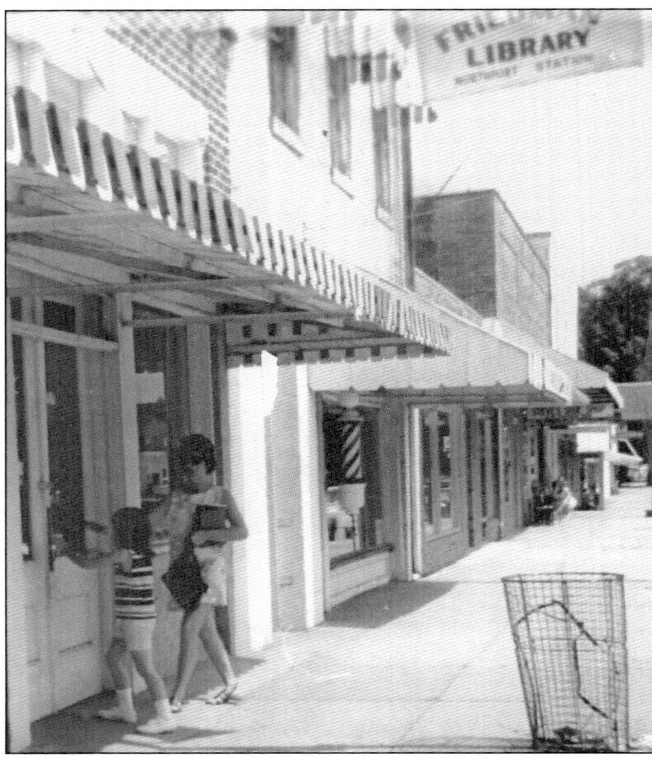

The Friedman Library, seen here in the early 1970s, provided Northport with a local branch library. The library started in 1955 in the Northport Community Center and in 1968, relocated to Main Avenue next to the barbershop. Caroline Dickson was the librarian for many years.

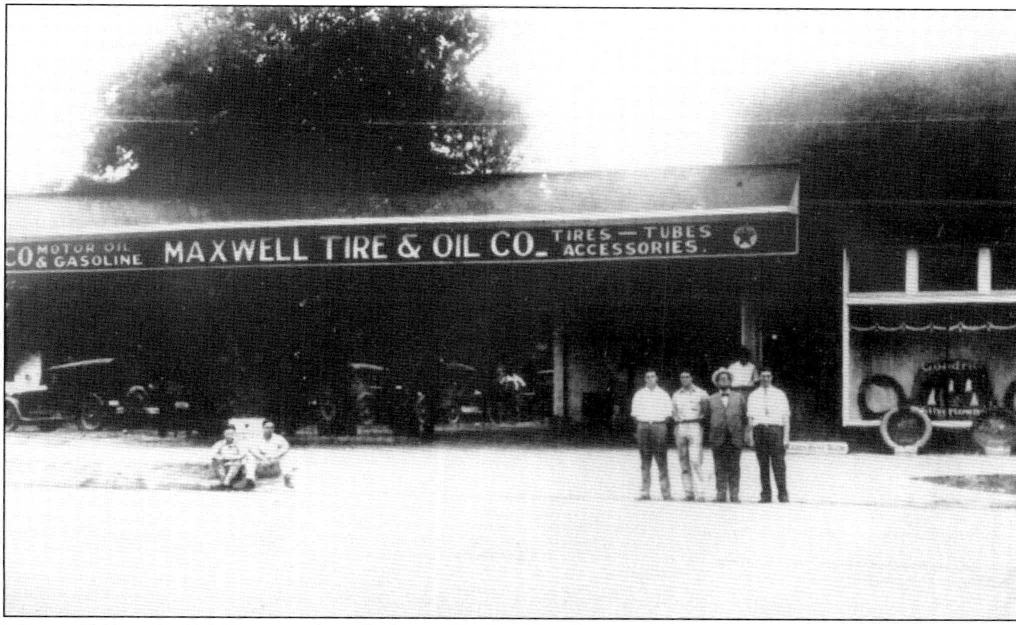

Maxwell Tire & Oil Co. was founded in 1922 by Charles R. Maxwell Jr. as a small service station in downtown Northport. He started the business as a way to pay for his school expenses as a University of Alabama ministerial student. Part of City Cafe and other retail shops now occupy the building on the west side of Main Avenue.

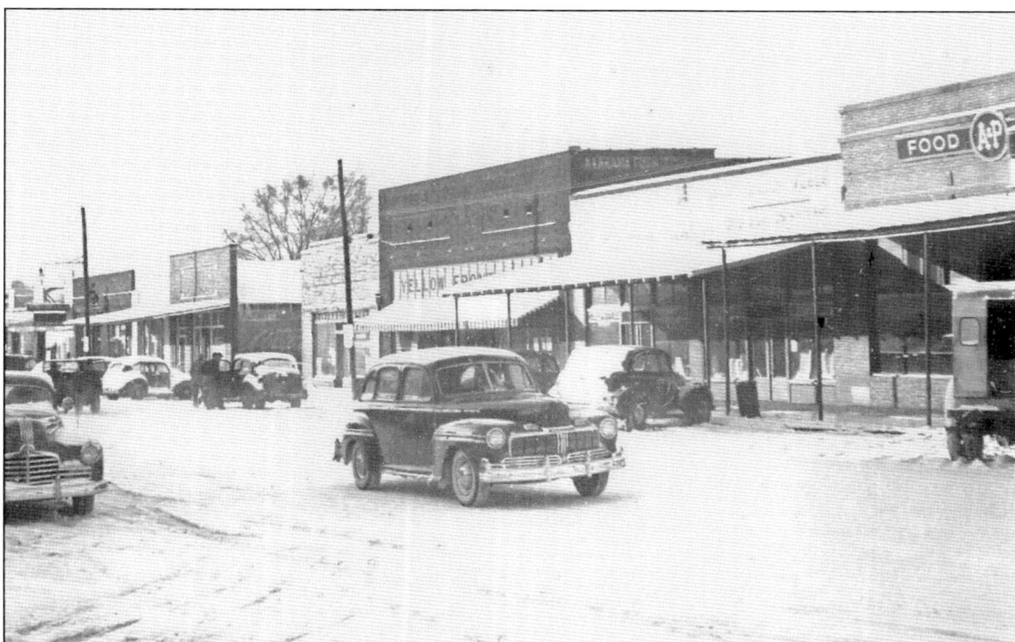

The Sumter Farm & Stock Company operated a chain of general merchandise stores in west and central Alabama called Yellow Front Stores. The Northport location was on the west side of Main Avenue next to the A&P. The stores were known for their "live and let live" prices. This photograph was taken around 1950.

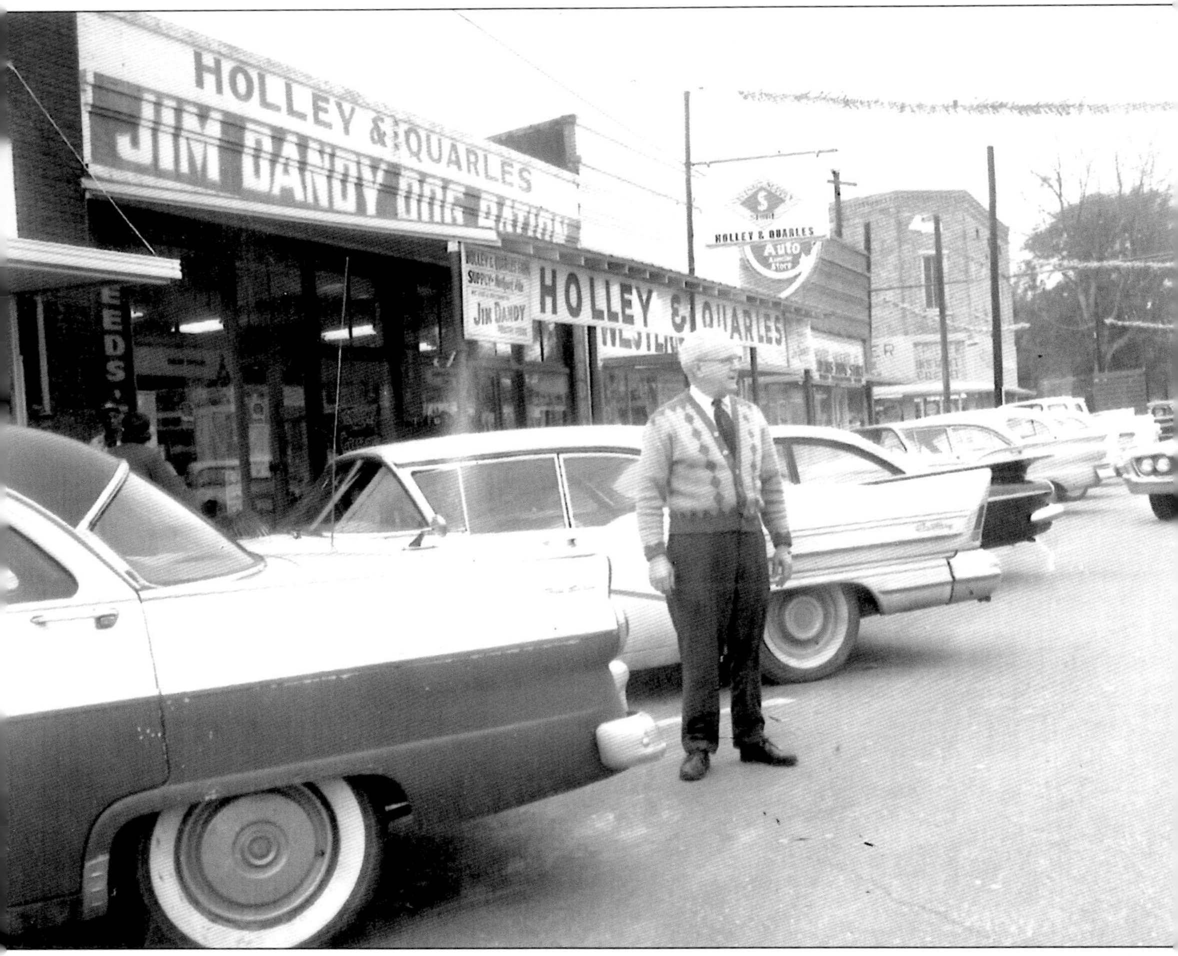

J.D. Quarles is seen here in front of the Holley & Quarles farm supply store on the west side of Main Avenue. Quarles co-owned and operated the business from 1947 until 1970. He served several terms on the city council and also served as mayor. This photograph was taken in the 1960s.

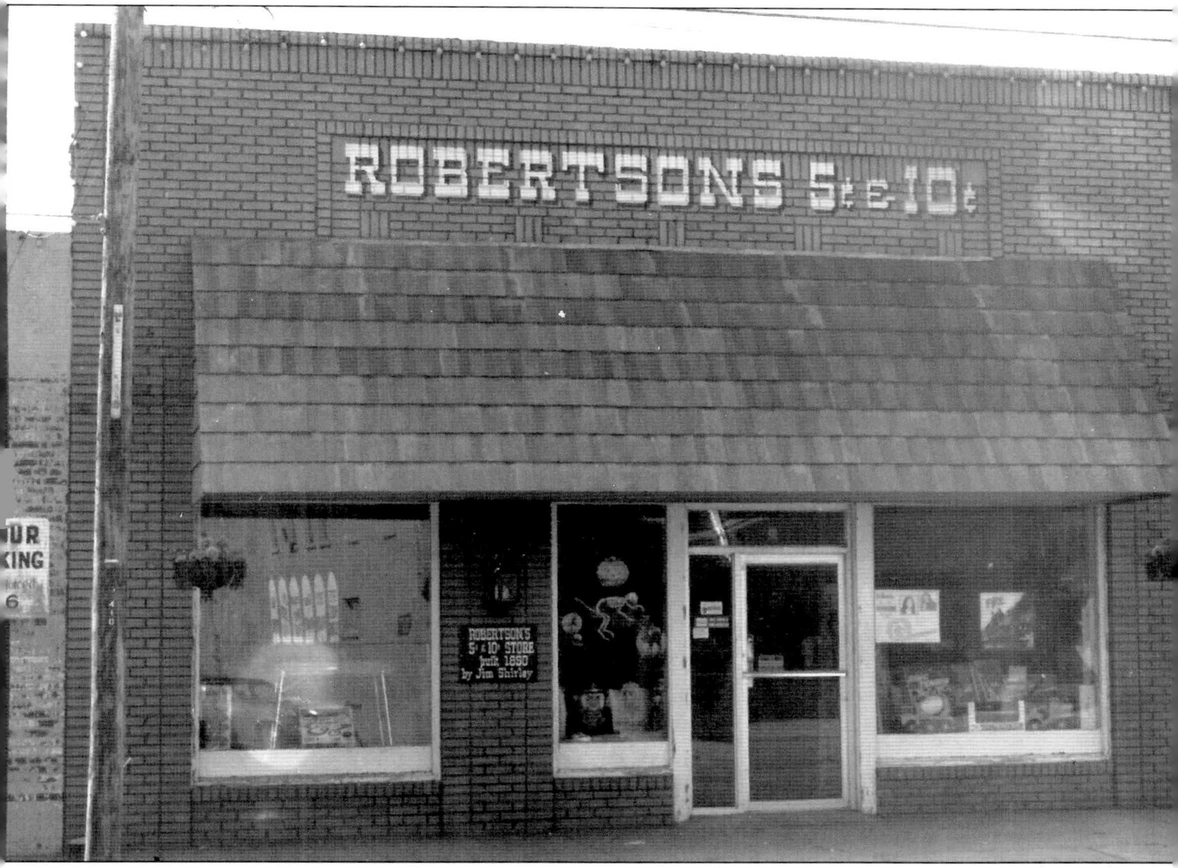

Robertson's 5 & 10 was located on the west side of Main Avenue in the 1850 Shirley Building. James Avery Adams and Earl Kelley also operated this location as a five and ten. In the 1940s, the Northport post office was located here. This heritage structure has been operated by Joe Hardy as Northport 5 & 10 since 1983.

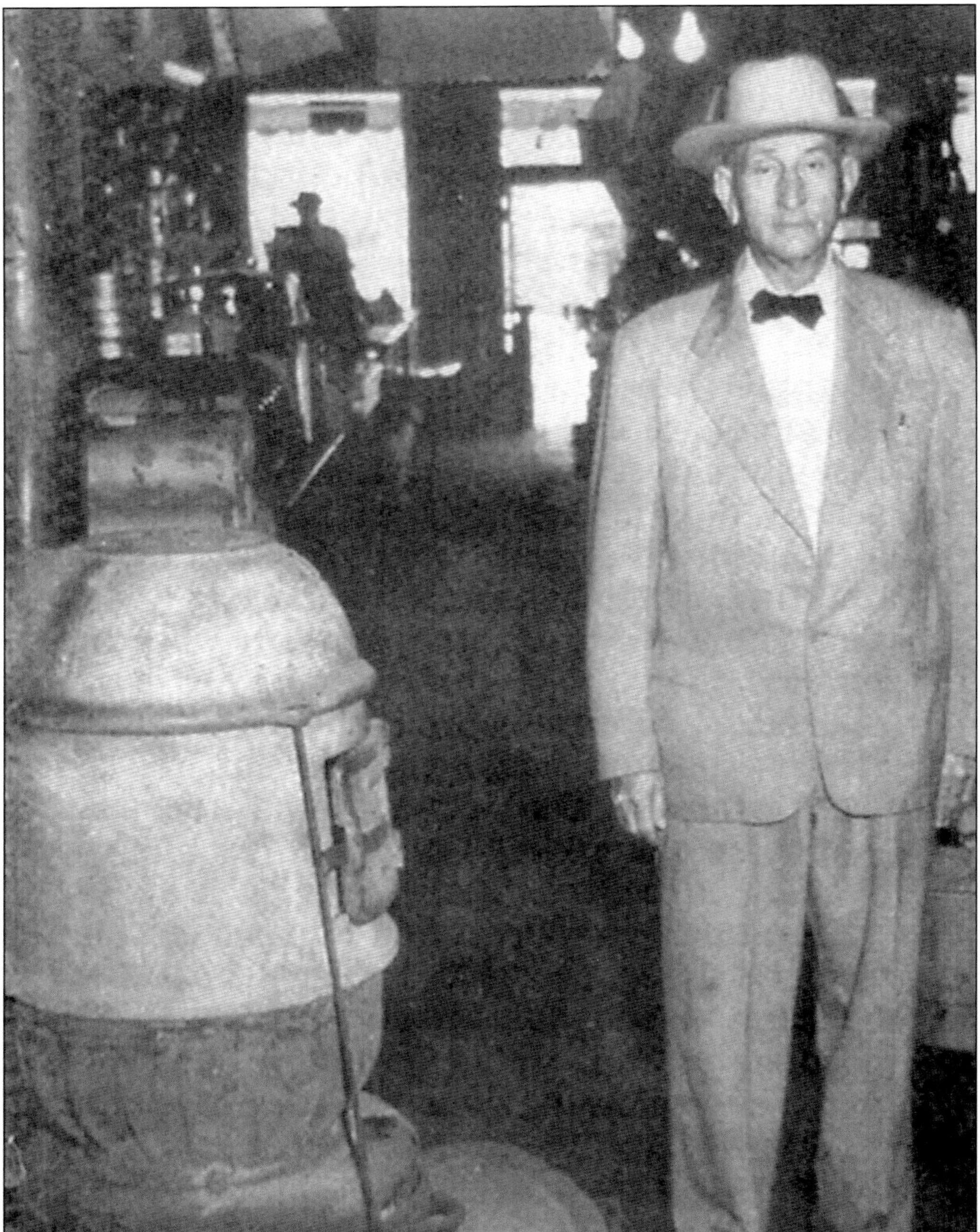

James Anders is pictured inside Anders Hardware in the 1940s. The potbelly stove seen here still sits in the store. Anders joined his two brothers, John and Lewis, in the family business in 1931 and later served the longest mayoral term in Northport's history, from 1938 to 1956. He guided the town during some difficult times, through the end of the Great Depression and several disastrous floods.

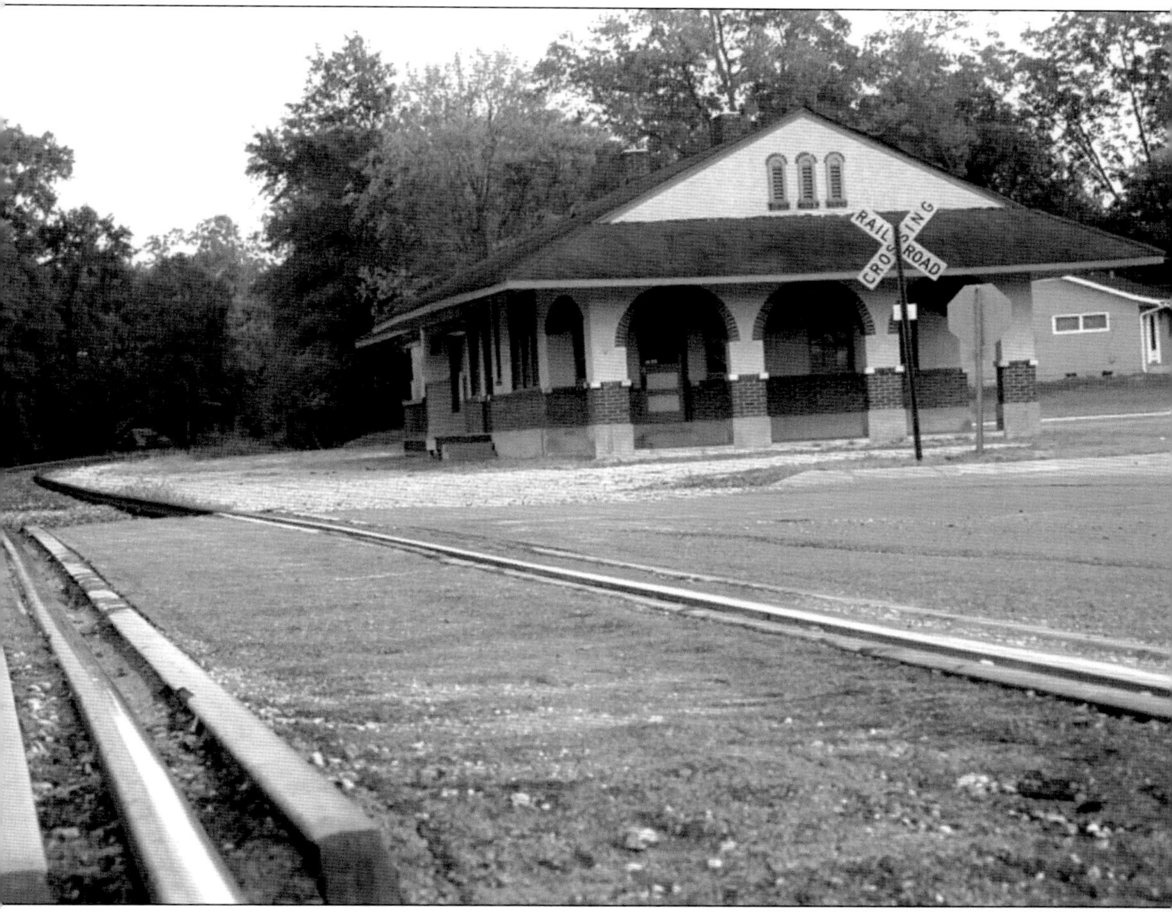

Today, the Northport railroad depot is a restored heritage structure owned by the City of Northport and operated as a historic site and model-train exhibit by the Black Warrior Model Rail Society. This building was used for both passenger and freight service by the Gulf, Mobile & Ohio Railroad. The depot remained in service until the late 1960s, when it was sold by the railroad company. It was later acquired by the city, and, with the help of a federal grant and hard work by the members of the Black Warrior Model Rail Society, was completely restored. It is the home of the Black Warrior Model Rail Society and one of Alabama's finest model-train displays.

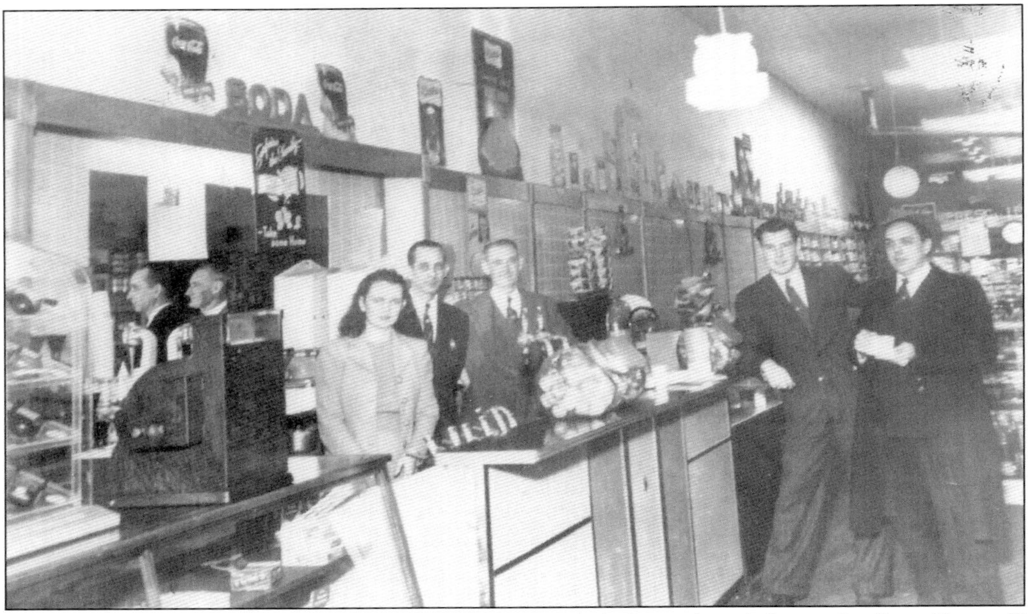

Seen in this interior view of Adams Drug Store around 1948 are, from left to right, Inez Chappell Shirley, Carl Adams, Norris Abels, store pharmacist Ray Jenkins, and Rev. John Rutland. Many locals remember the soda fountains at Adams Drug Store. Among the many "soda jerks" who worked there were Sonny Booth, Sam Faucett, and T.C. Cole.

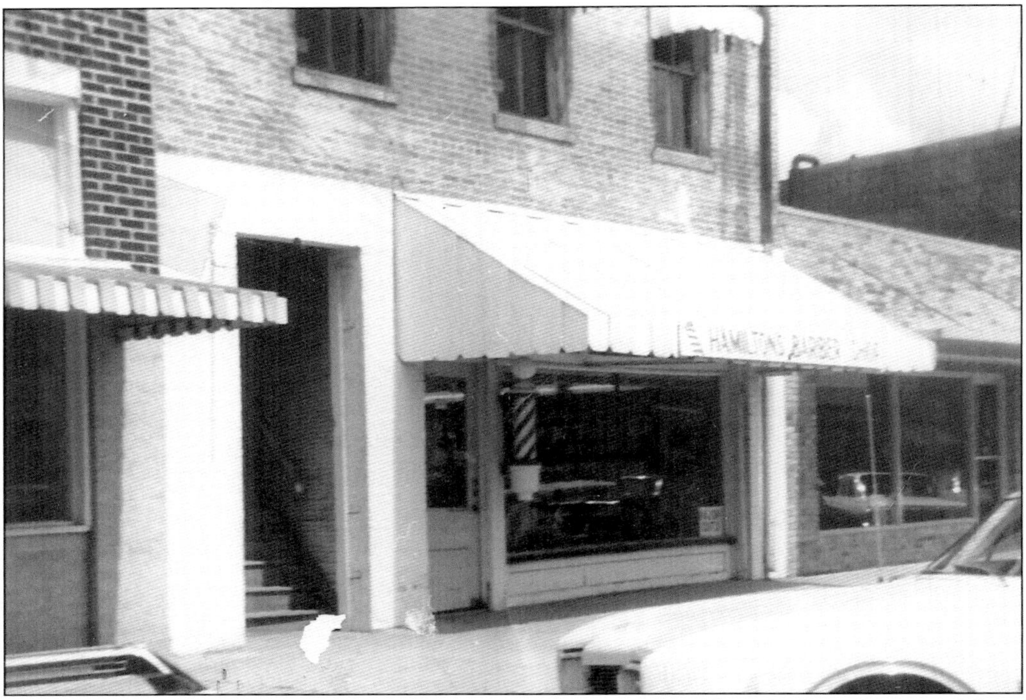

This 1970s photograph shows the Hamilton Barber Shop and the old city hall on the second floor. The Rice Bros. office and Anders Hardware are to the right. These commercial structures were built after a massive fire in 1891 destroyed much of the east side of Main Avenue.

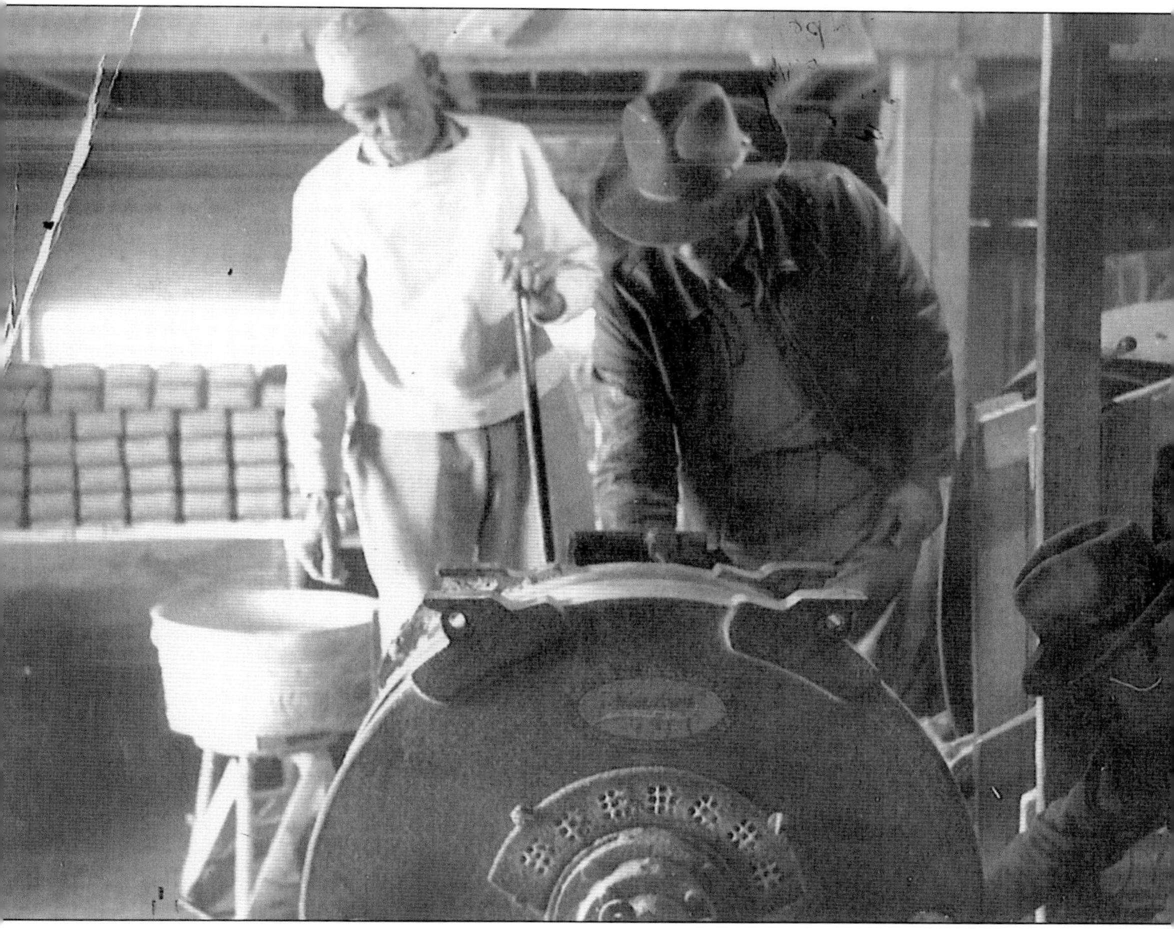

J.D. Payne (in black hat), pictured here in the 1950s, was the owner of the Northport gristmill, on Fourth Street just west of Main Avenue, which was one of the first mills to use electrical power. After grain was harvested and the chaff was removed, it was called grist and was ready to be ground by the miller. Water-powered mills, such as the Hargrove Mill in Northport, were eventually replaced by electric mills.

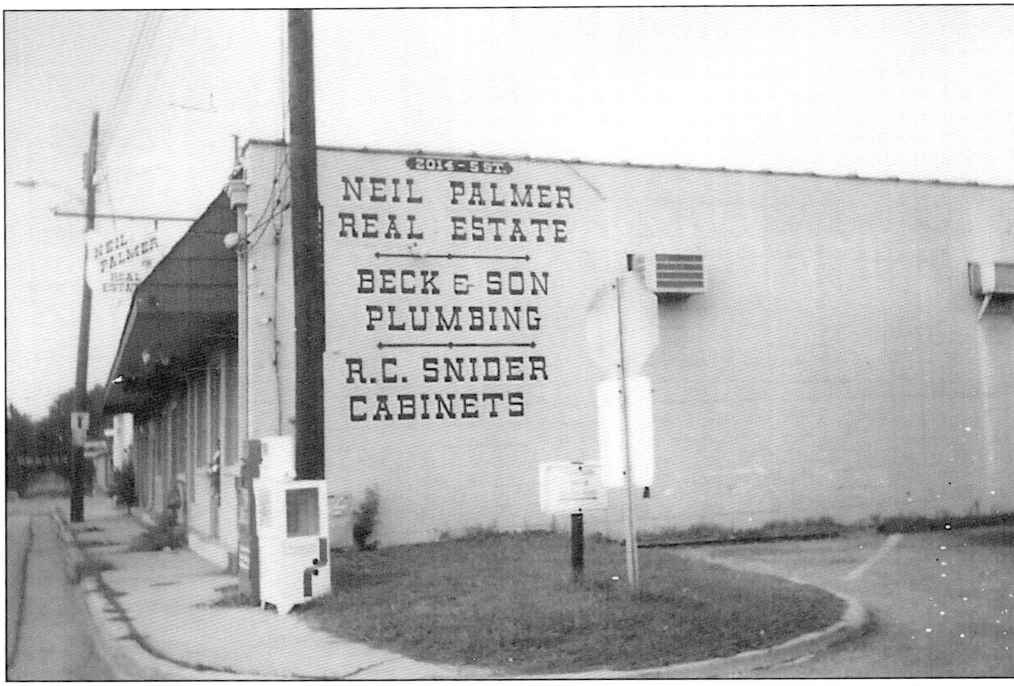

Three businesses—Neil Palmer Real Estate, Beck & Son Plumbing, and R.C. Snider Cabinets—operated from the former location of Josh Palmer's blacksmith shop, to the east of the current Kentuck Museum. Neil Palmer was Josh Palmer's son. This photograph was taken in the mid-1960s.

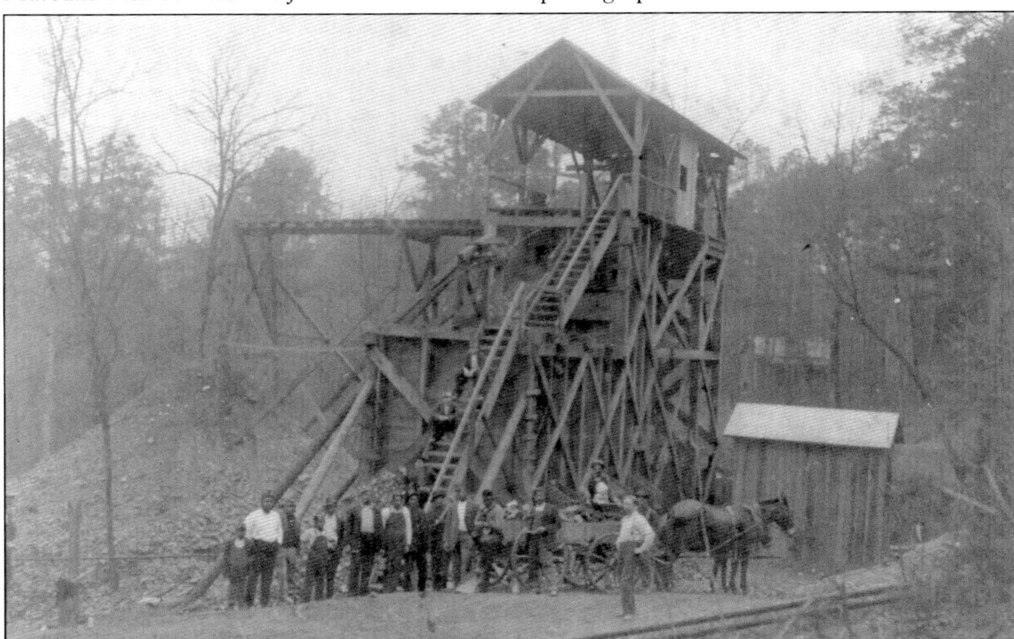

Coal was discovered in the mid-1800s on the north side of the river, a short distance east of Northport. The road to Rice's coal mine from downtown Northport was later named Rice Mine Road. This deep-well-shaft coal mine was near the current site of Cypress Inn Restaurant.

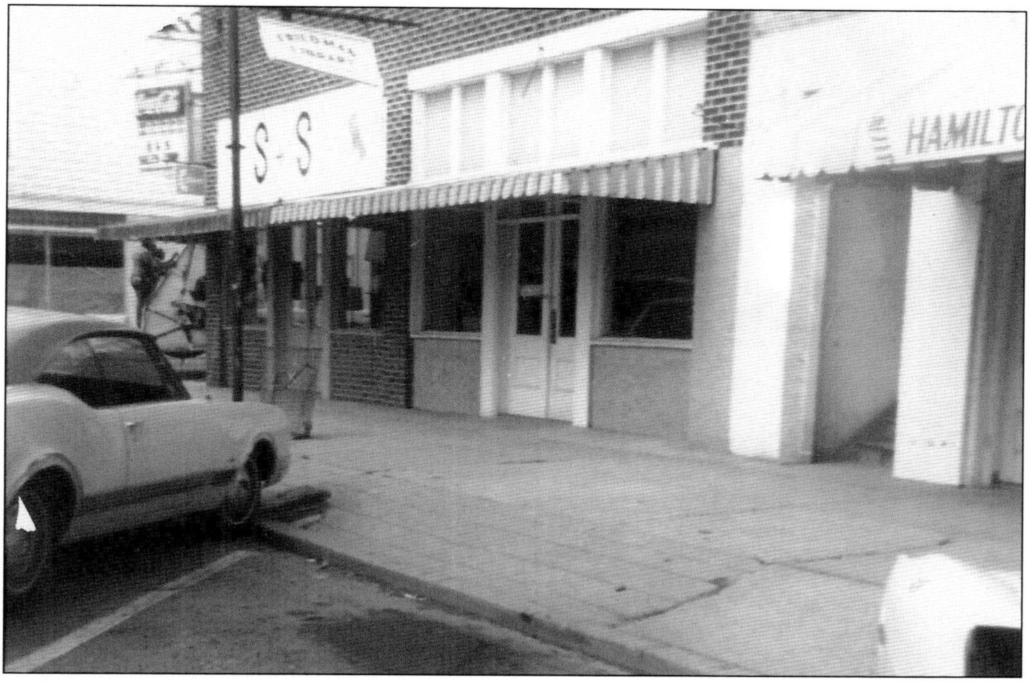

In the 1970s, The southeast corner of Columbus Street (now Fifth Street) and Main Avenue was home to the S&S Shop, the Friedman Library's Northport branch, the upstairs entrance to the old city hall, and Hamilton's Barber Shop. This photograph shows the condition of the sidewalks and streets along Main Avenue. The revitalization efforts of the "Northport Renaissance" in the 1990s led to vast improvements to downtown Northport.

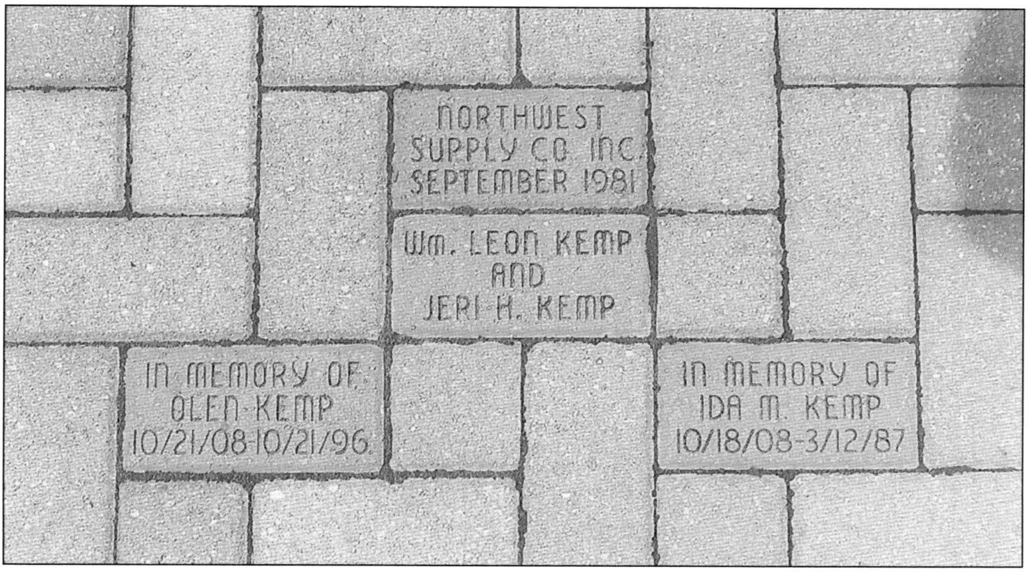

Today, engraved brick pavers line the streets of downtown Northport. Sold as a fundraiser for the Northport Renaissance efforts in the mid-1990s, these heritage bricks honor and memorialize many of Northport's citizens and also provide historical information. Also, installed on each light pole is a historic fact about a current or former building or event that occurred near that pole.

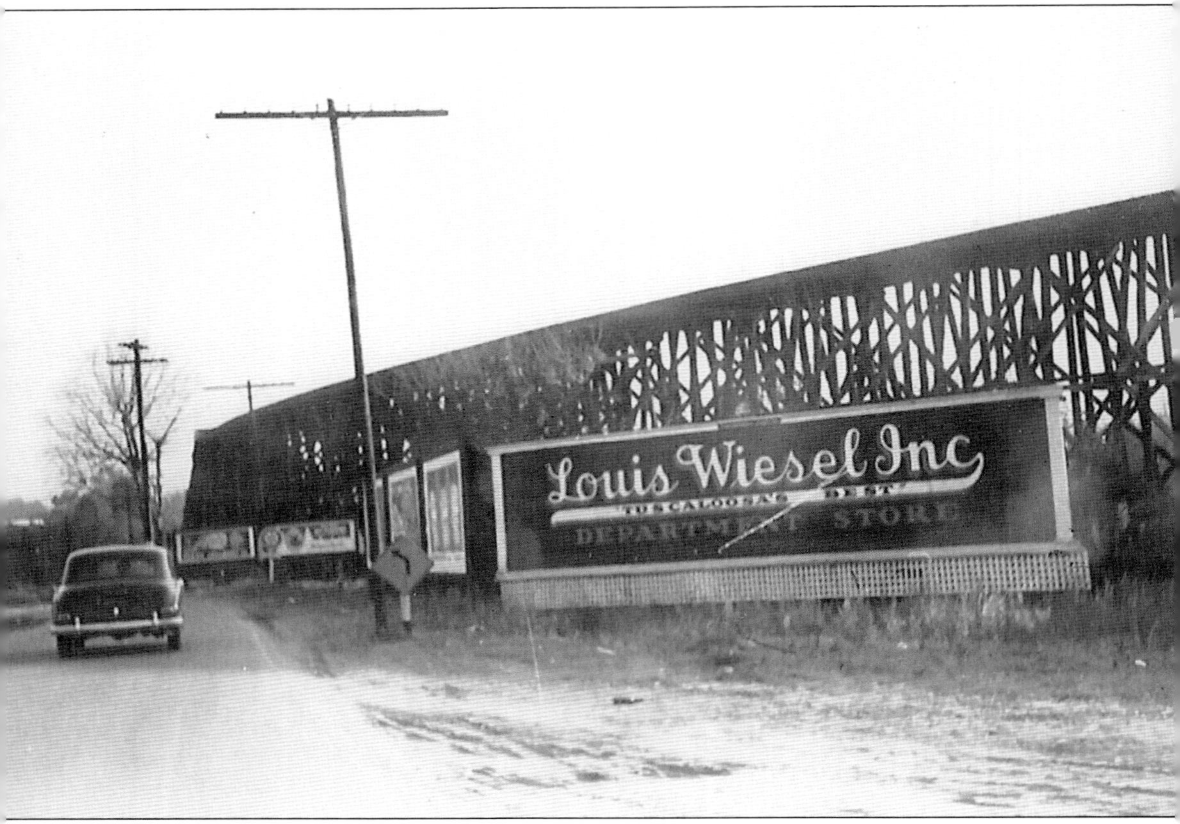

This 1948 photograph shows the Gulf, Mobile & Ohio railroad trestle at the corner of Main Avenue and First Street in downtown Northport. Many locals remember the billboard promoting Louis Wiesel, Inc., that sat next to the train trestle. Wiesel established his women's department store in Tuscaloosa in the late 1920s.

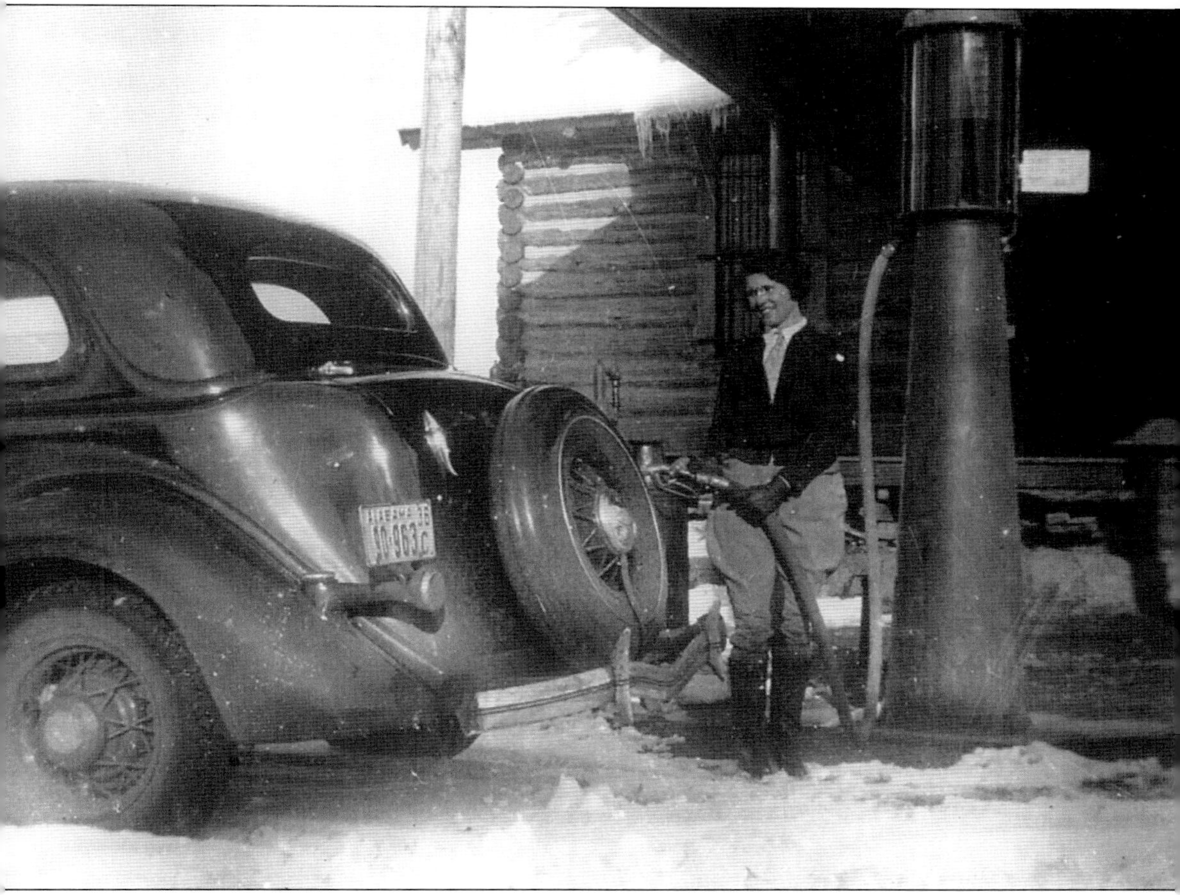

Pictured is Grace Davis as she pumps gas at her father, O.M. Davis's, gas station on Bridge Avenue in Northport. The car is a 1935 Ford that has a 1936 tag. Up to 10 gallons of gas could be pumped by hand into the glass container at the top. Gravity emptied the gas into the car's tank. There is snow on the ground and icicles hang from the roof of the station. (Courtesy Bruce Davis and the Tuscaloosa Area Virtual Museum.)

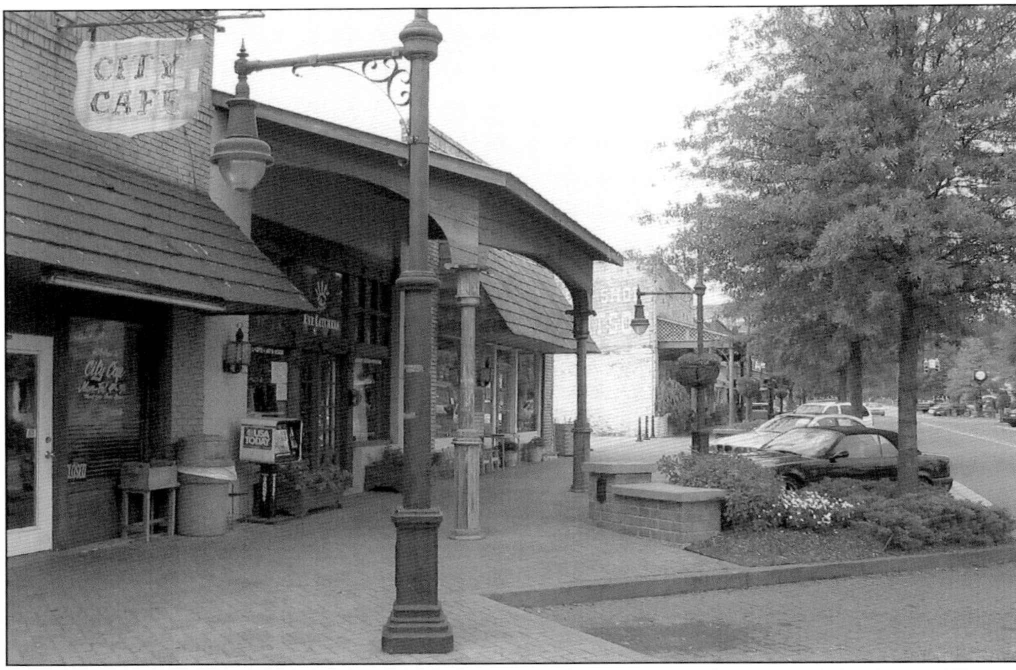

City Café is one of the oldest continuously operating businesses in downtown Northport. Known as Joe's Café before World War II, it was operated by several Northport citizens over the years, including Jo Knight, Trannie Barnett, Hoyt Brasher and Bill Hitt, and Joe Barger. "Miss Trannie," as Barnett was often called, was well known for her chicken potpie, which she made and served at City Café.

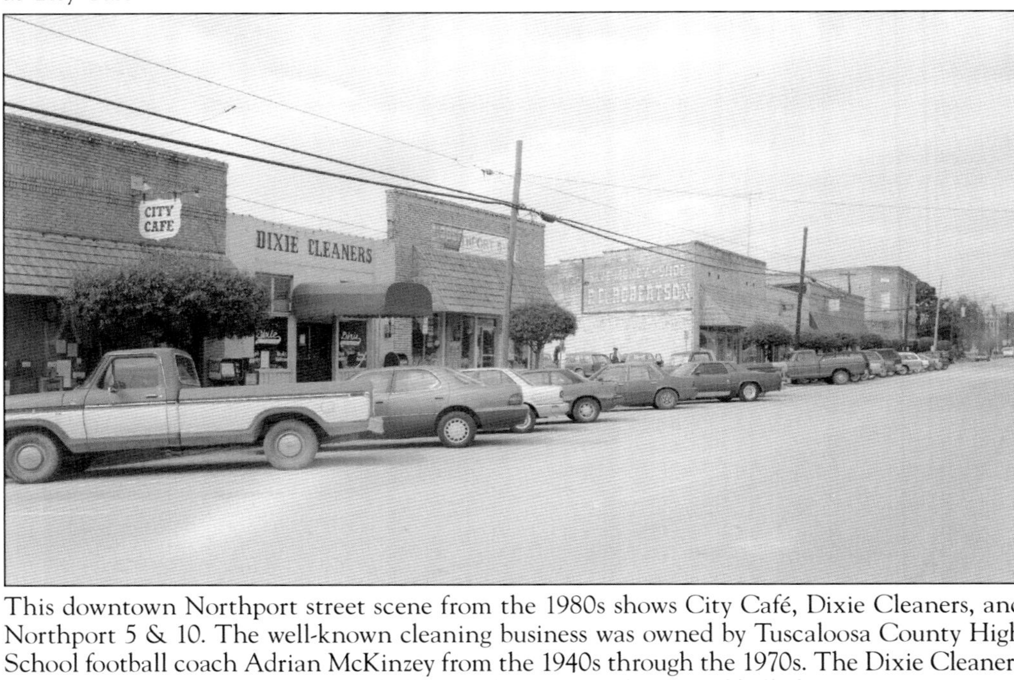

This downtown Northport street scene from the 1980s shows City Café, Dixie Cleaners, and Northport 5 & 10. The well-known cleaning business was owned by Tuscaloosa County High School football coach Adrian McKinzey from the 1940s through the 1970s. The Dixie Cleaners Softball Club was a city-league softball team during Tuscaloosa's softball glory years.

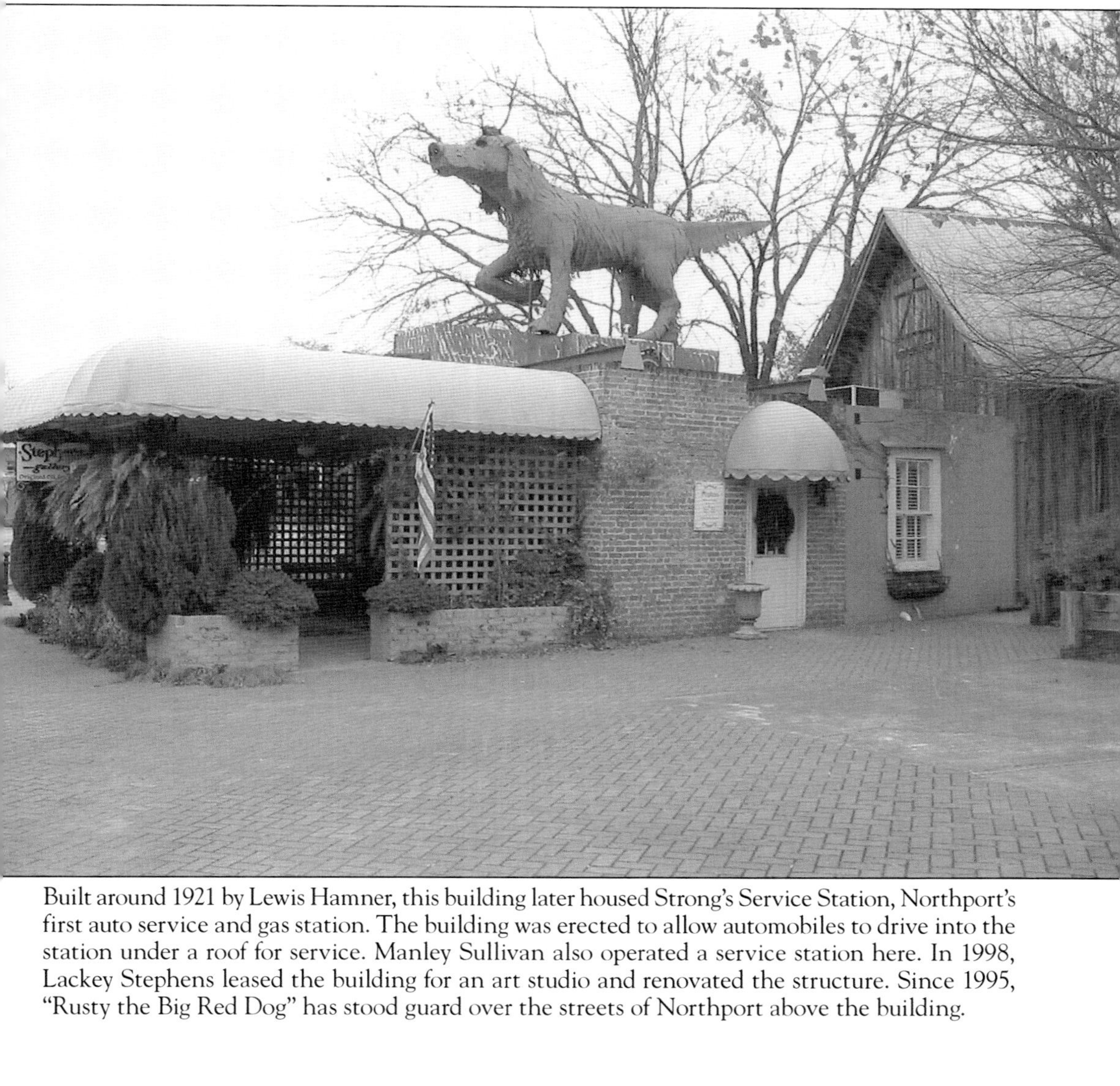

Built around 1921 by Lewis Hamner, this building later housed Strong's Service Station, Northport's first auto service and gas station. The building was erected to allow automobiles to drive into the station under a roof for service. Manley Sullivan also operated a service station here. In 1998, Lackey Stephens leased the building for an art studio and renovated the structure. Since 1995, "Rusty the Big Red Dog" has stood guard over the streets of Northport above the building.

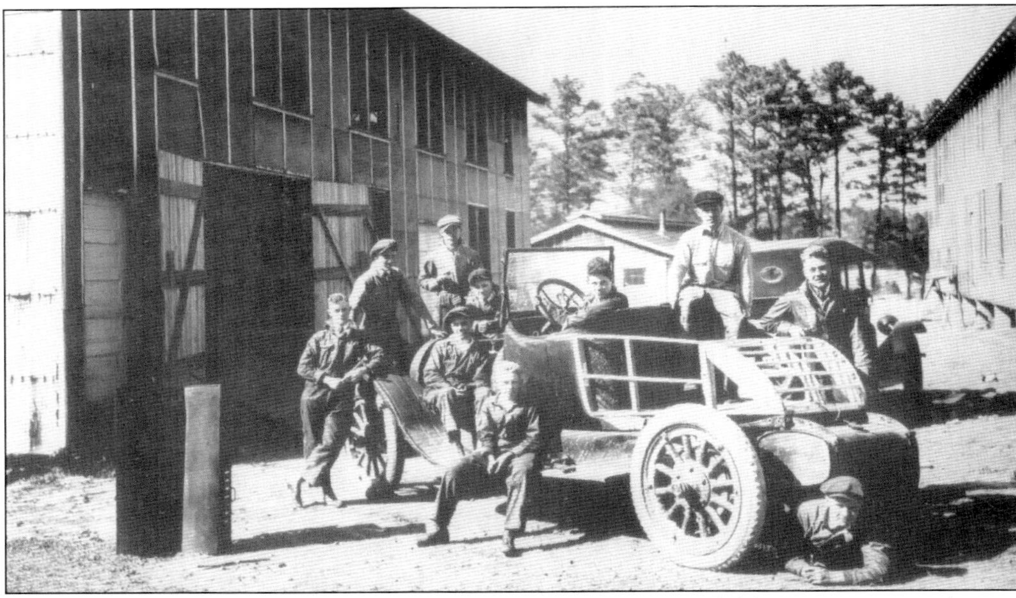

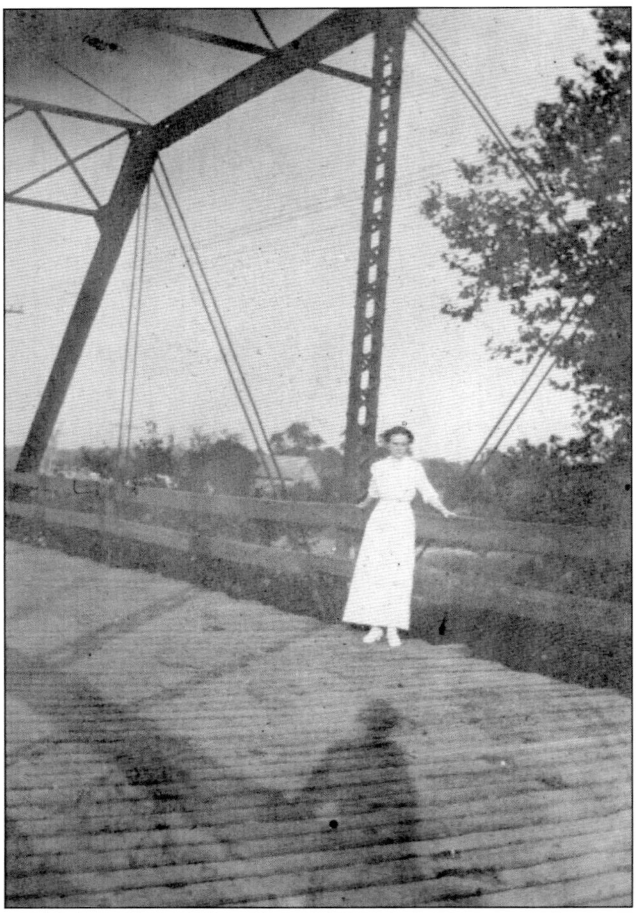

Winfield Scott Persinger, the son of Elias Persinger, an early Northport settler and merchant, sits at the wheel of the first automobile in Northport in 1905. W.S. Persinger was issued a license, signed by probate judge James C. Brown at the county courthouse, in February 1904. At the time, a state license was not required. This photograph was taken at Persinger's boathouse and boat launch, on the river west of the train trestle.

Edna Adams Anders stands on the old land bridge on First Street in 1916. The land bridge was between the Barnes & Norris cotton gin and crossed a small creek. The creek was covered and the bridge was removed in the 1930s. Sage Olmstead operated a hat factory at this site during the Civil War.

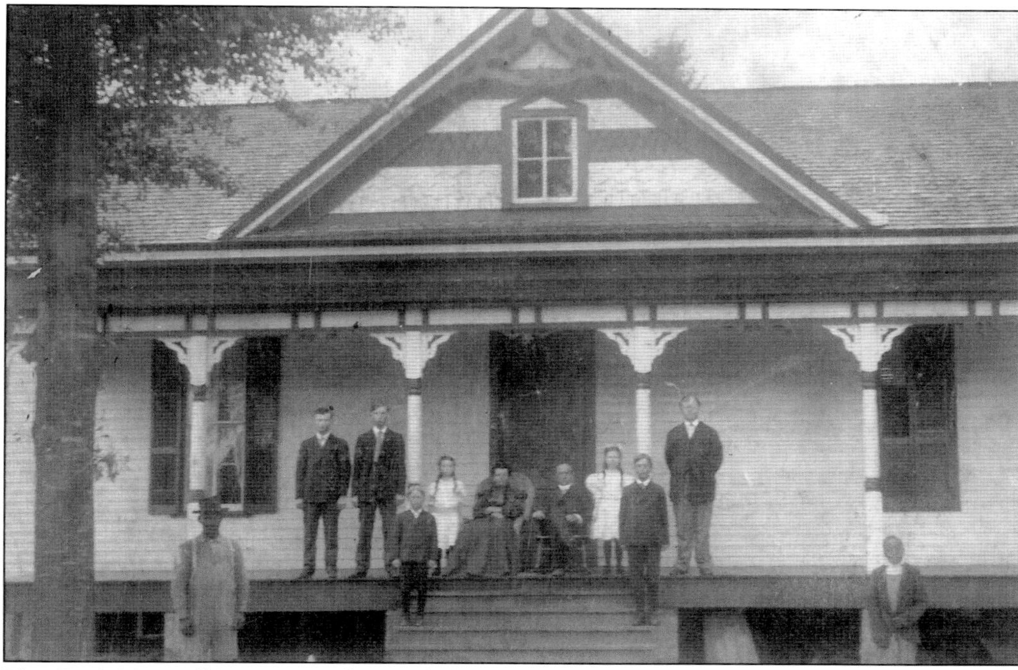

The Cook-Maxwell House, seen here, is behind Northport Methodist Church on Seventh Street. It was built in the 1830s by Robert Cook, who helped establish the Academy School and also donated the land to build the church. Standing on the front porch is the family of Nathaniel G. Holley, who lived here for several years.

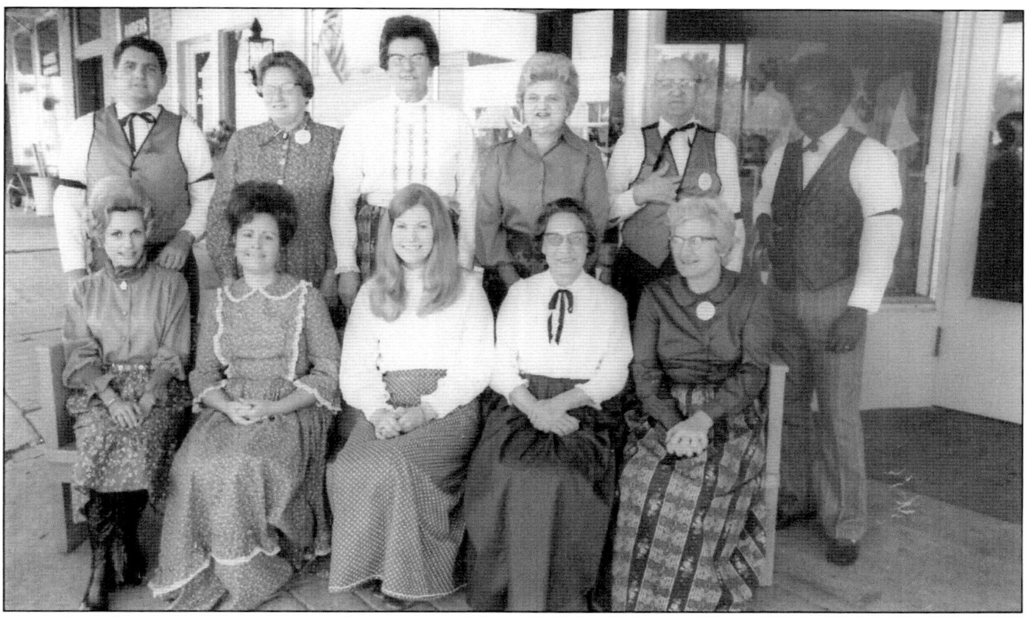

This photograph of the Faucett's Department Store staff was taken during the 1971 Northport Centennial. The staff includes, from left to right, (first row) Jan Faucett, Betty Dunn, Janice Howard, Necie Robertson, and Mattie Phillips; (second row) John Paul Faucett, Irene Maxwell, Opal Culbert, Gwendolyn Faucett, Bill Faucett, and Willie King.

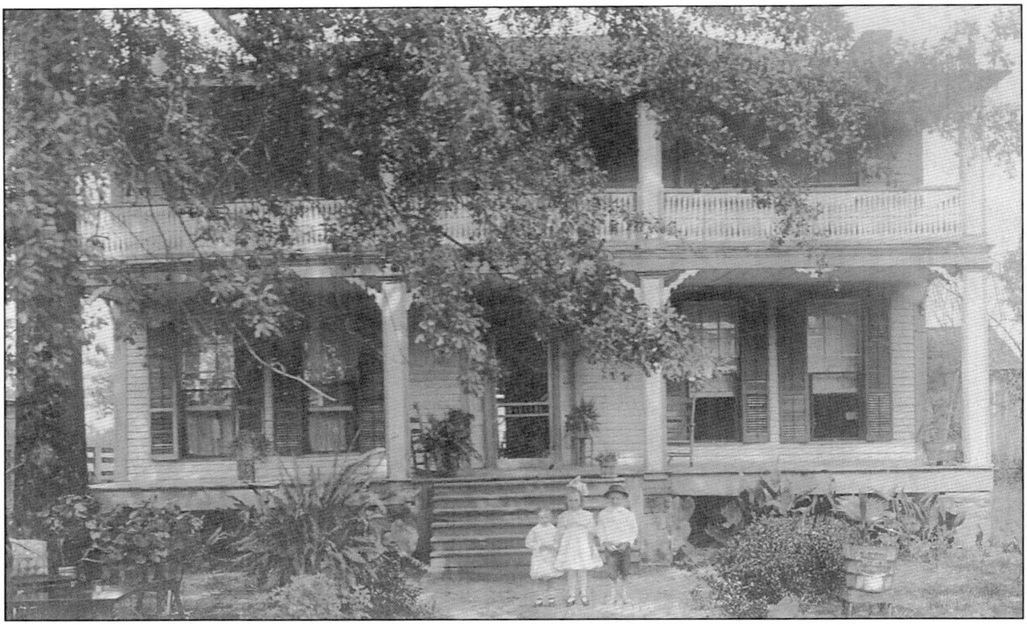

This antebellum home south of Northport Methodist Church on Main Avenue was built for Elias Persinger by Alfred Ray in the mid-1830s. It was destroyed by the 1932 tornado. In front of the house, from left to right, are children Louis Anders, Josephine Anders, and Bill Faucett around 1910.

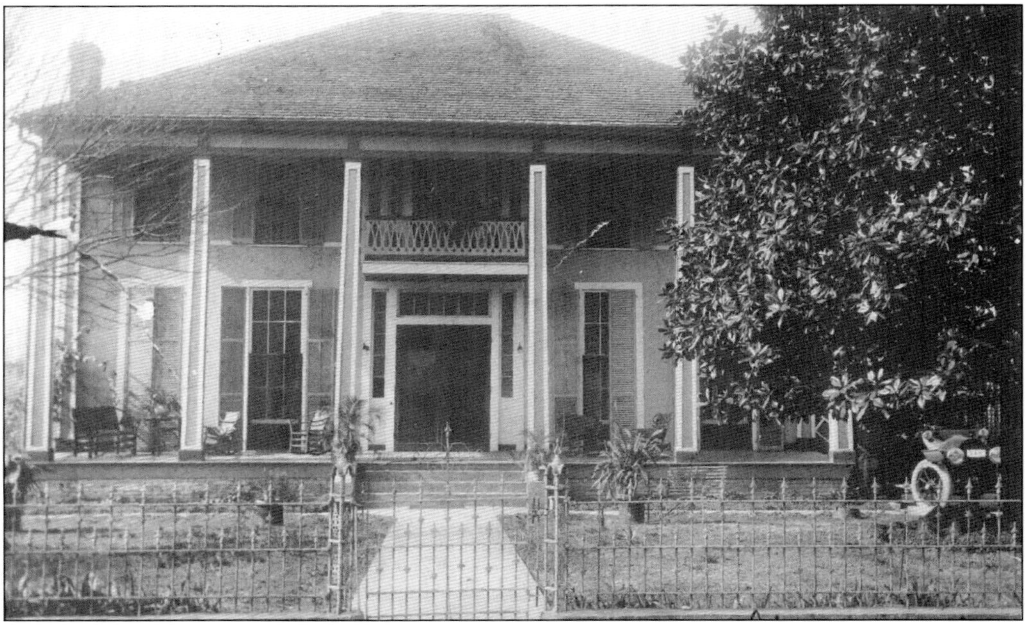

This home stood next to the Northport Methodist Church on Main Avenue. It was built for Jesse Van Hoose prior to the Civil War and was later remodeled into a two-story home by Thomas Franklin Rice. The home was later owned by George W. Christian Sr.; in its last years, it became an apartment building. Today, Crowning Glory Beauty Salon occupies the lot. The magnolia tree that stood in front of the home was well known in the town.

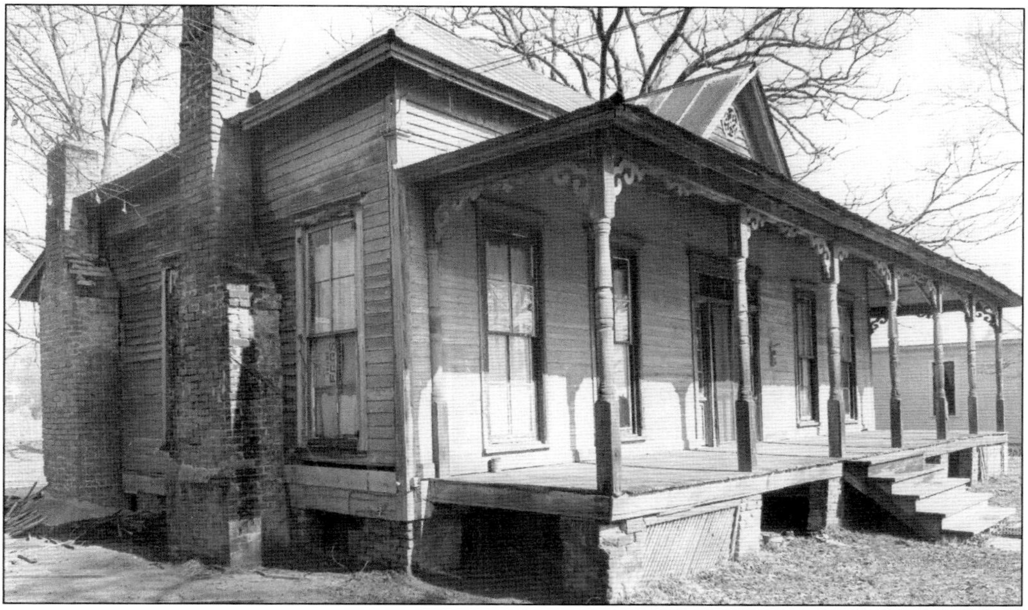

Built about 1850, this home, on Fifth Street across from Williamson Cemetery, was first owned by Dr. T.S. Danviss and later by Joseph Shirley. It was built in several stages over the years. The original building included two front rooms separated by a hall, with plaster walls and chimneys in each room. The house eventually had three handmade exterior brick chimneys. A second room and side porch were added, and later, two other rooms were added. The floors in the original part of the house were wide-width heart pine. The structure was lost to a fire in 2000.

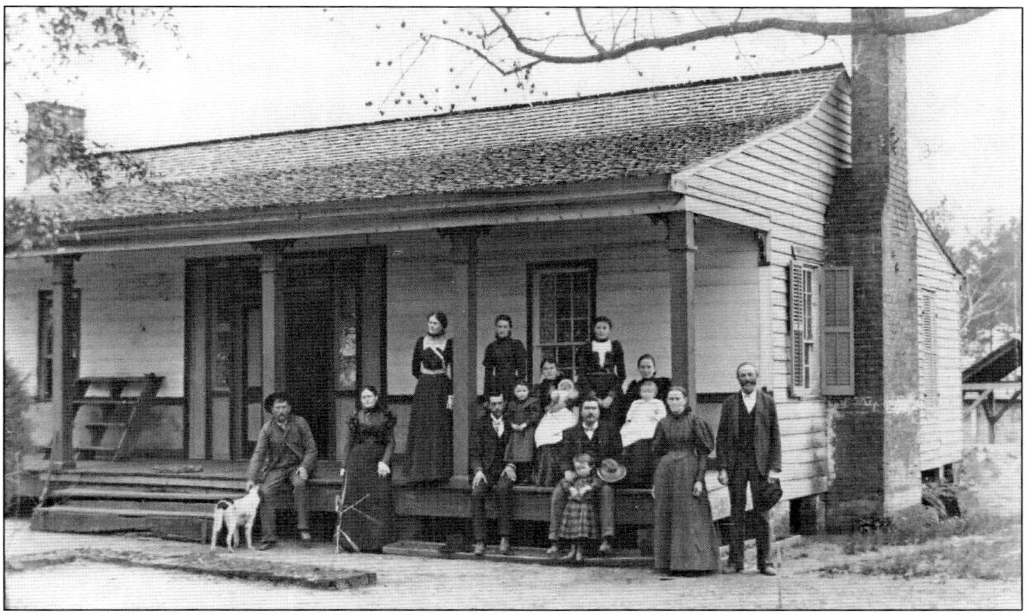

This 1896 photograph was taken in front of the 1825 Findley-Stone house on Main Avenue just north of downtown. Civil War veteran Washington Deason, his wife, Frances A. Bell Deason, and their family lived in the house for several years around 1900. The house was originally built with wooden pegs and featured an open dogtrot and two exterior brick chimneys.

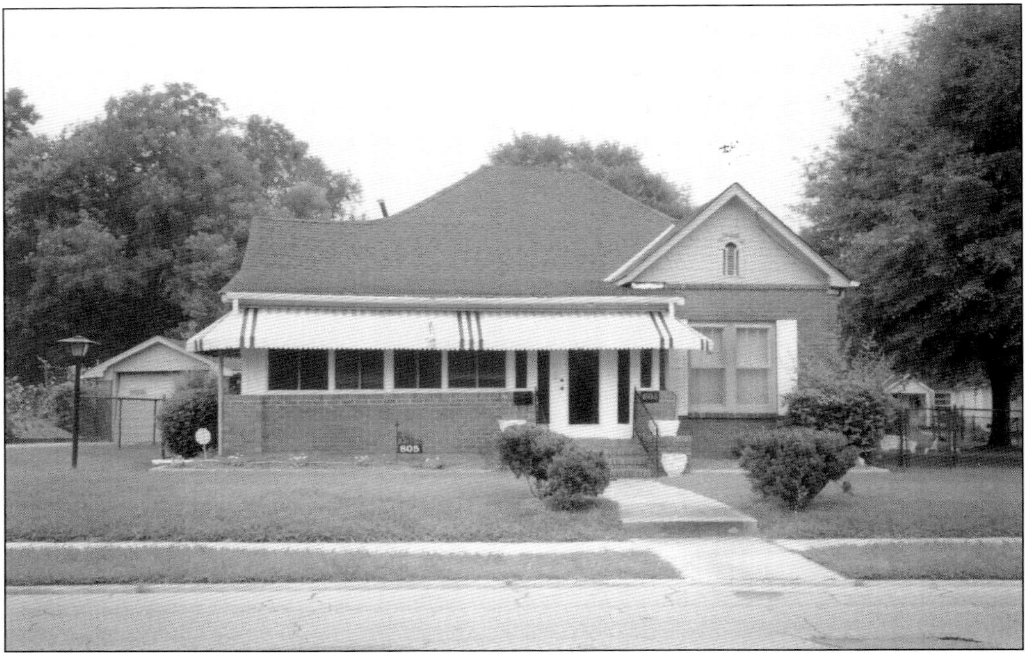

John Green and his wife, Bama, lived in this home at 805 Bridge Street (later Bridge Avenue) before 1900. Their oldest daughter, Cleola Green Ivery, and her husband, Richard Ivery, were given the property after Bama's death. The original wooden structure has been remodeled with brick, although interesting wood detail still remains on the front gable.

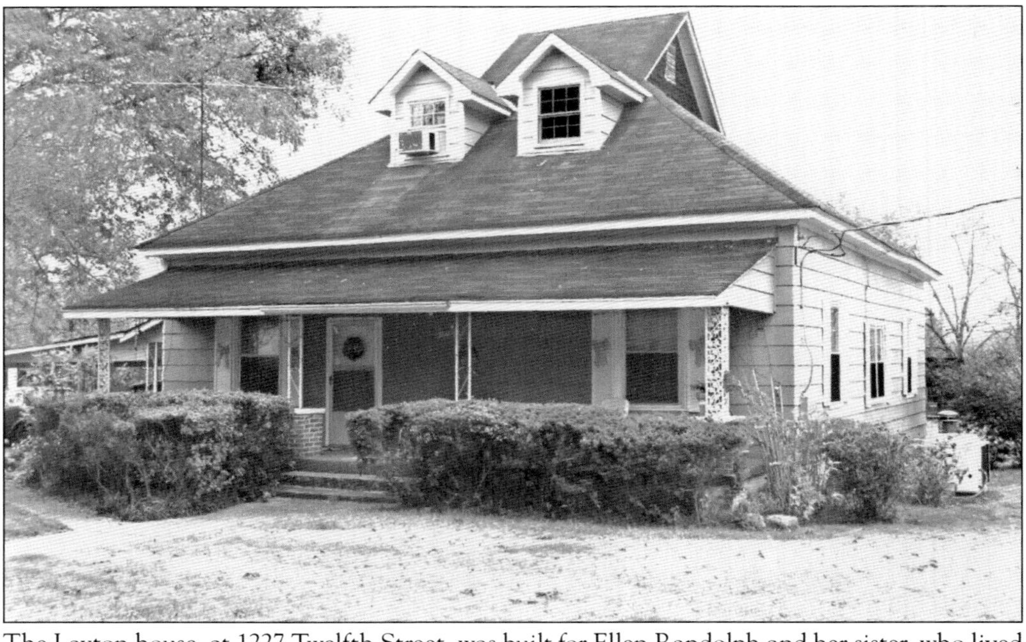

The Layton house, at 1227 Twelfth Street, was built for Ellen Randolph and her sister, who lived there for several years. Later, Vonnie Layton and his wife, Willie, acquired the property and raised 10 children in the home. Now surrounded by many newer dwellings, it once stood alone as one of the first houses built east of old Bridge Avenue.

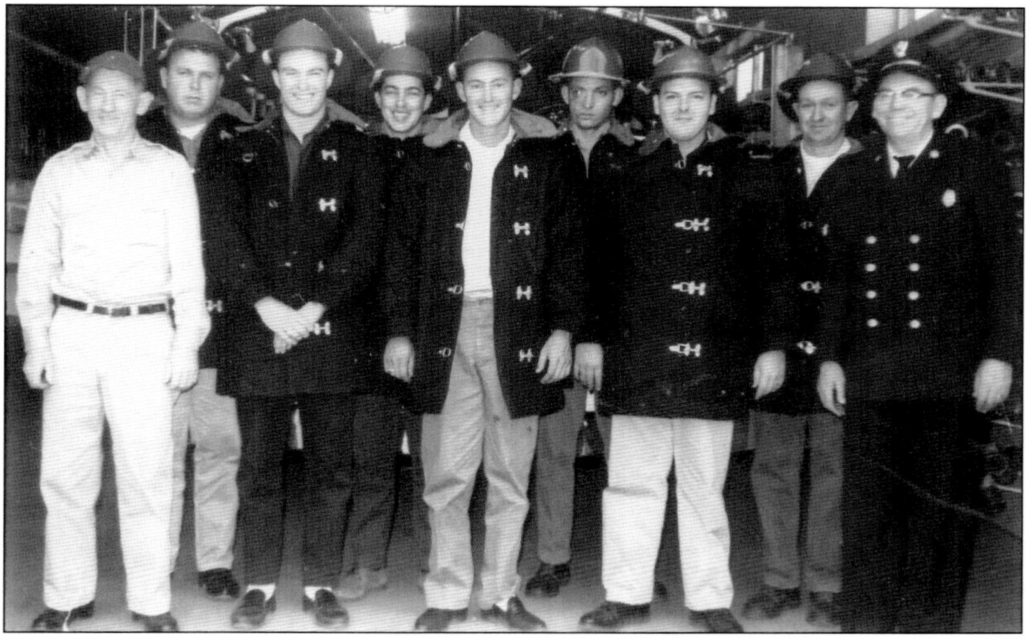

The first firefighters of the Northport Fire Department are seen here in 1965. From left to right are Assistant Chief Woolsey Morris, Joe Christian, Jimmie Tittle, Roy Kelly, M.L. Lynn, Ray Hamner, Ray Bigham, Lonnie Fowler, and Chief H.C. "Happy" Hulgan.

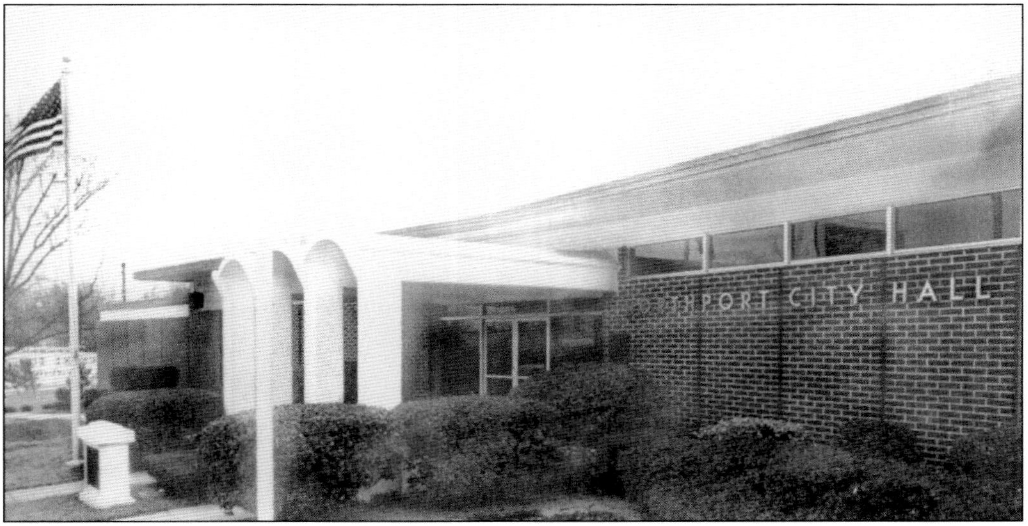

Northport's city hall moved into this new building in 1965, which housed it until 1985, when McCrory Village on McFarland Boulevard was converted into a new city hall and civic center.

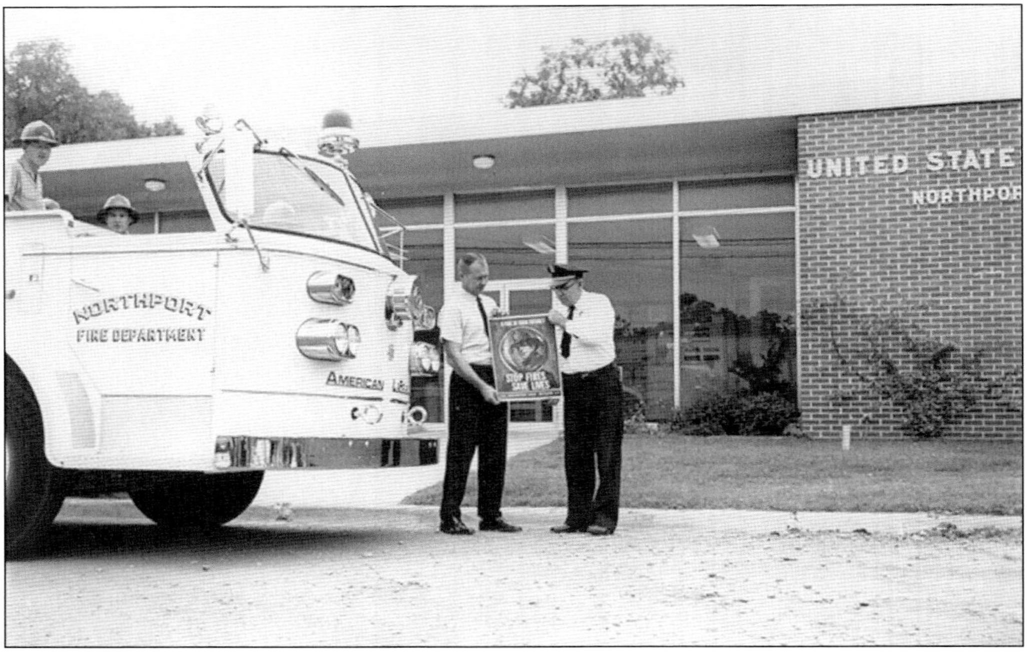

One of the first fire trucks purchased by the Northport Fire Department is seen here in front of the post office on Fifth Street. Pictured are Northport Post Office employee Sonny Booth (left) and Fire Chief H.C. "Happy" Hulgan.

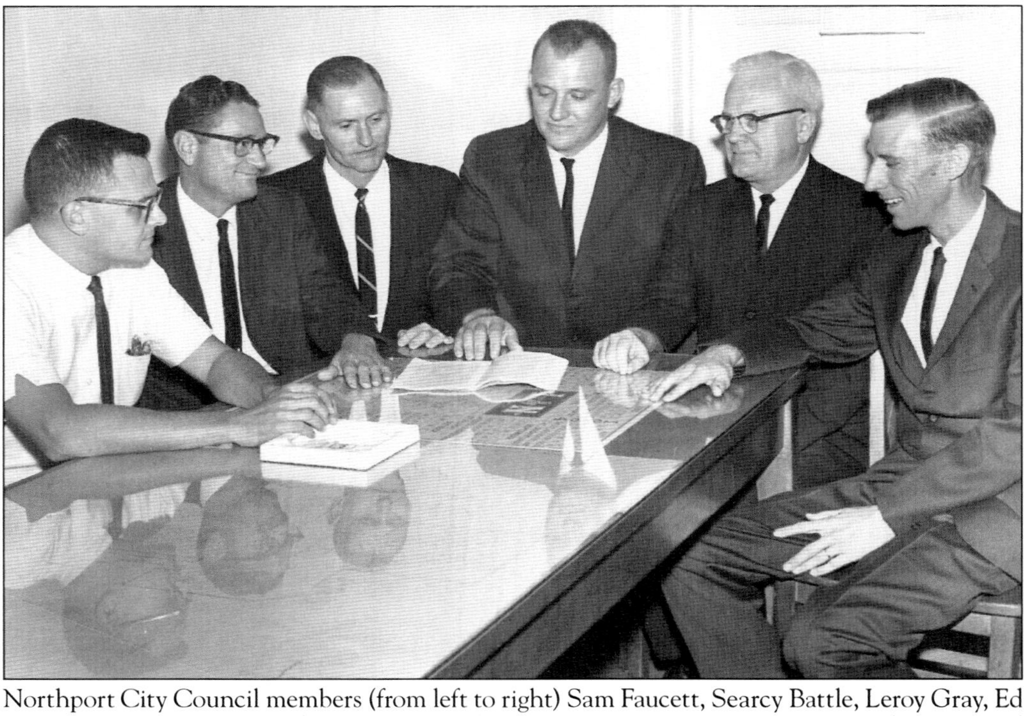

Northport City Council members (from left to right) Sam Faucett, Searcy Battle, Leroy Gray, Ed Robertson, J.D. Quarles, and Dan Morrison discuss city matters in 1964. Faucett, Robertson, and Morrison would later serve as mayors of Northport.

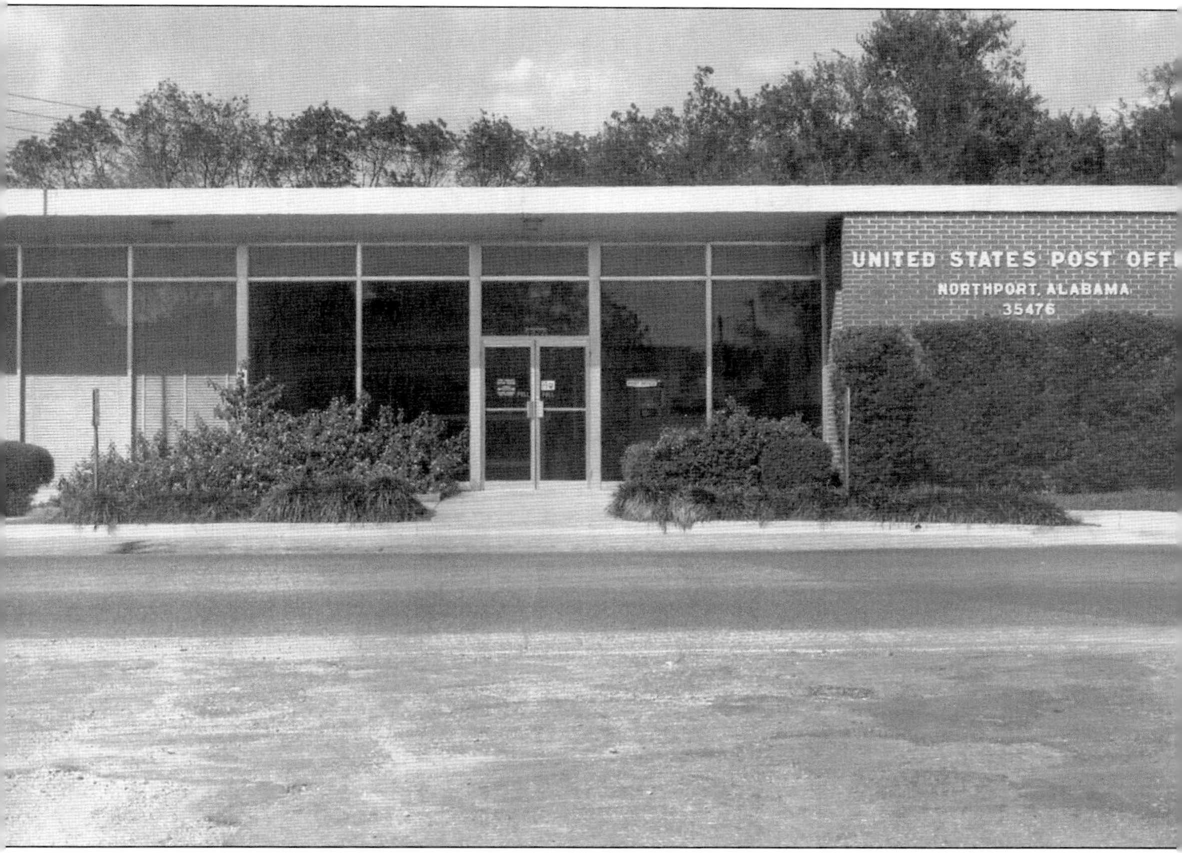

The Northport post office on Fifth Street is seen here. A new post office was built in 1980 on McFarland Boulevard. This building is now owned by the Kentuck Association and includes art studios and meeting spaces. It was named the Georgine Clarke Building in honor of the organization's first executive director.

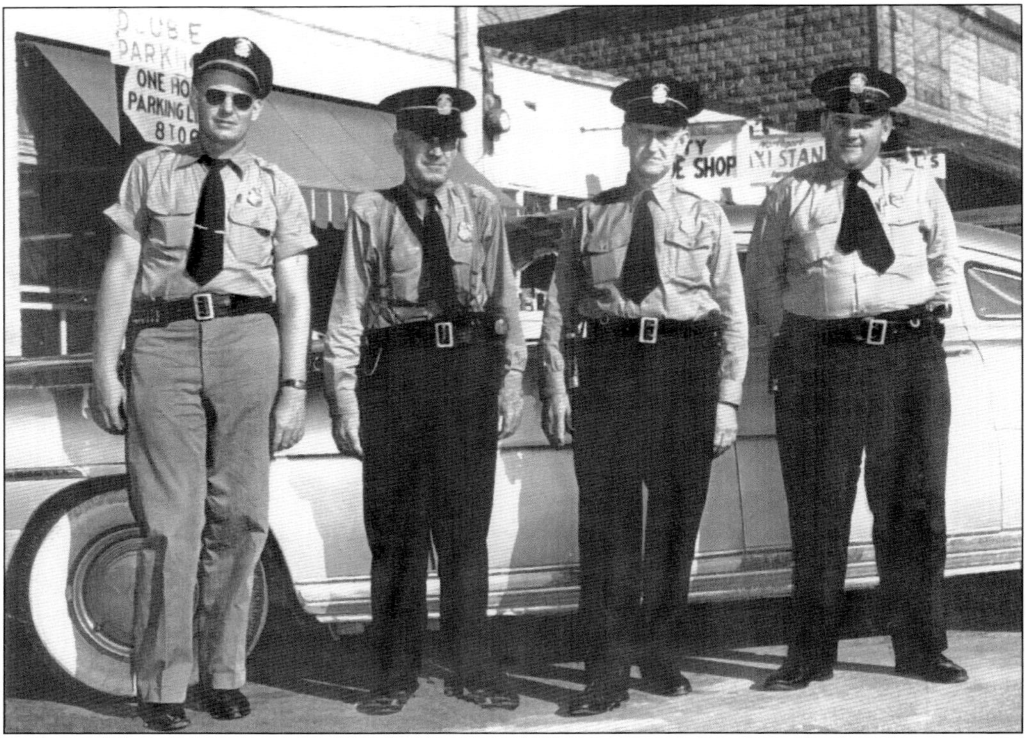

From left to right, Officer J.L. "Red" Bodiford, Officer Arthur A. Strickand, Officer C.S. Gann, and Chief Theron Dyer of the Northport Police Department pose in front of a police cruiser in 1949. Officer Bodiford was a motorcycle policeman, and Officer Gann walked a beat. Behind them are the City Shoe Shop and the Northport Taxi Stand.

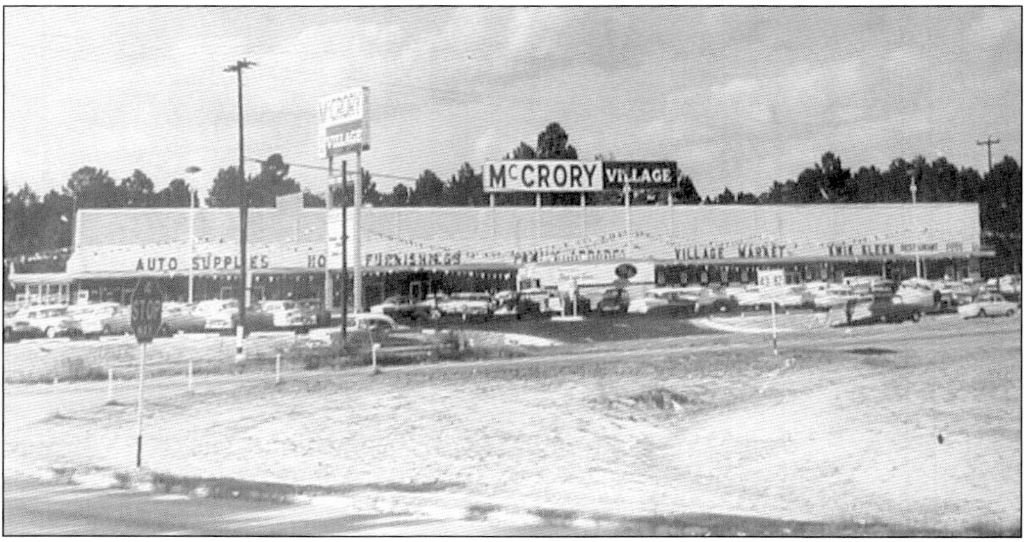

This 1962 photograph shows the newly built McCrory Village on McFarland Boulevard, which served as a shopping center until 1985 when it was converted into the Northport City Hall and Civic Center. In 2010, the structure received a $2.5 million renovation that promises city hall many more years of use in the same location.

In 1979, Sue Prewitt opened Sue's Flowers at 405 Main Avenue. Over the years, that location has housed several businesses, including the Northport Bakery, Furniture Auction Mart, Nichols Restaurant, and the Dinner Bell Restaurant. In the early 1900s, a livery stable was located at the site.

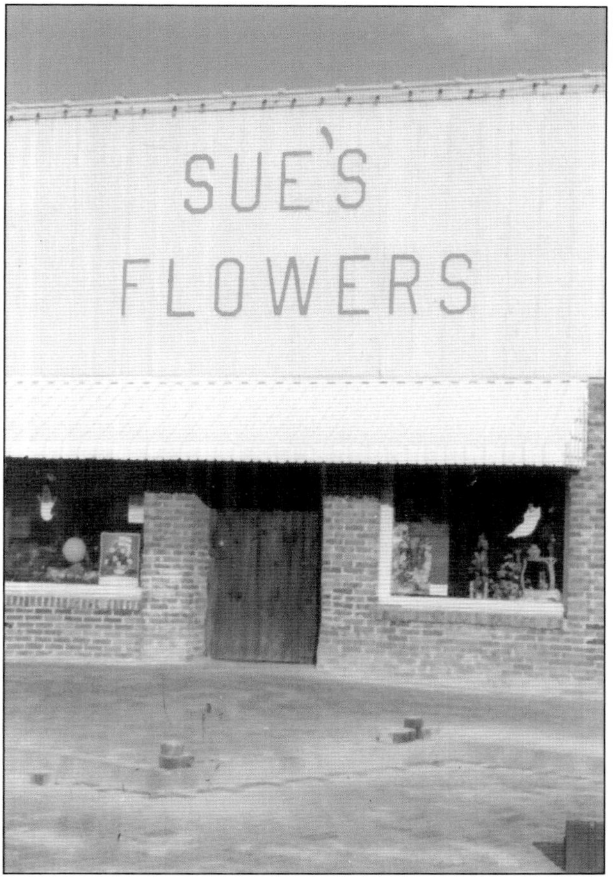

Now the Kentuck Museum, the structure below on Main Avenue was built around 1920 to replace a 100-foot-deep structure that burned in 1908. The upper level was designed as a residence for the storeowner, which was typical of the era. In the 1940s, Mr. and Mrs. T.A. Cole operated Cole's Cleaners from this location. In the 1950s, Sonny Booth and Joe White purchased the business from the Coles and operated B&W Cleaners here. The Coles lived upstairs for many years. Later, the Northport Chamber of Commerce was located here. The Kentuck Museum Association has occupied the building since 1980.

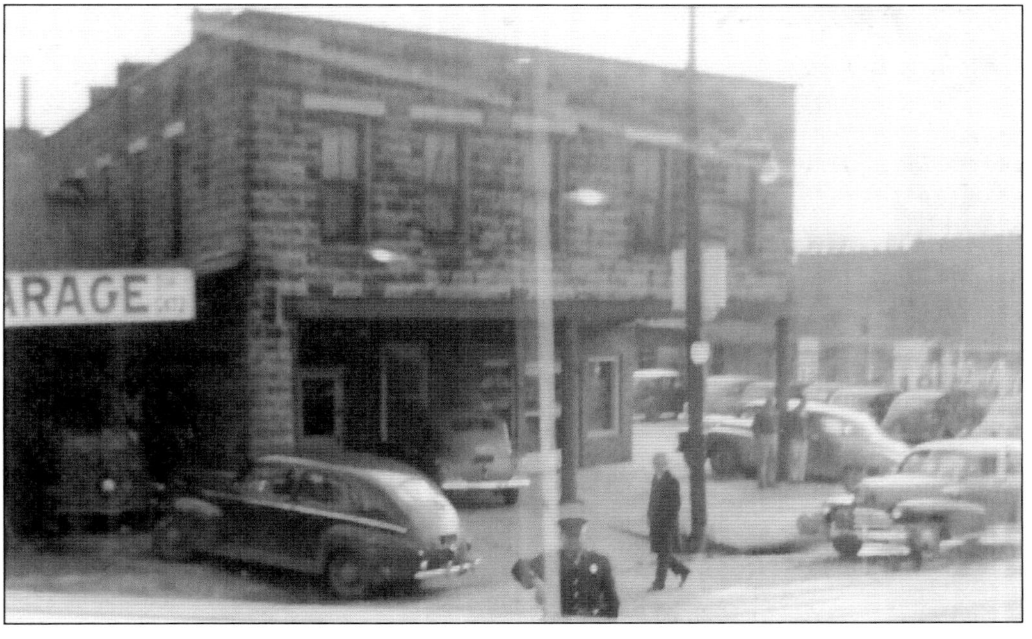

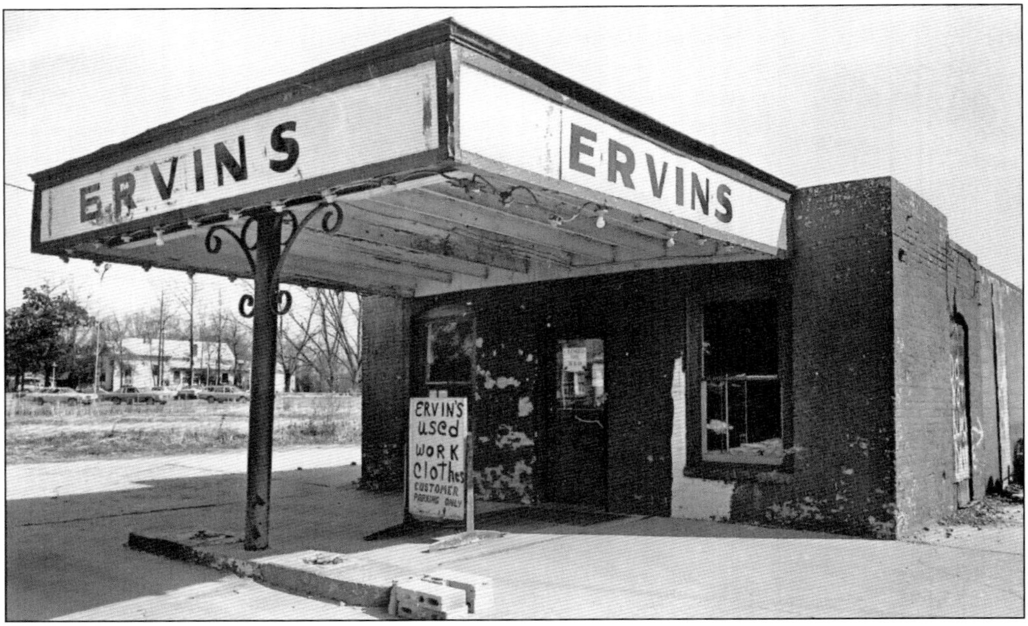

Through the 1970s, Ervin's Used Work Clothes occupied the former Strong's Service Station building at 509 Main Avenue. The business, operated by Ervin and Polly Koon from 1964 to 1999, eventually changed its name to Ervin's Work & Western Wear. Stephens Studio & Gallery occupies the building today.

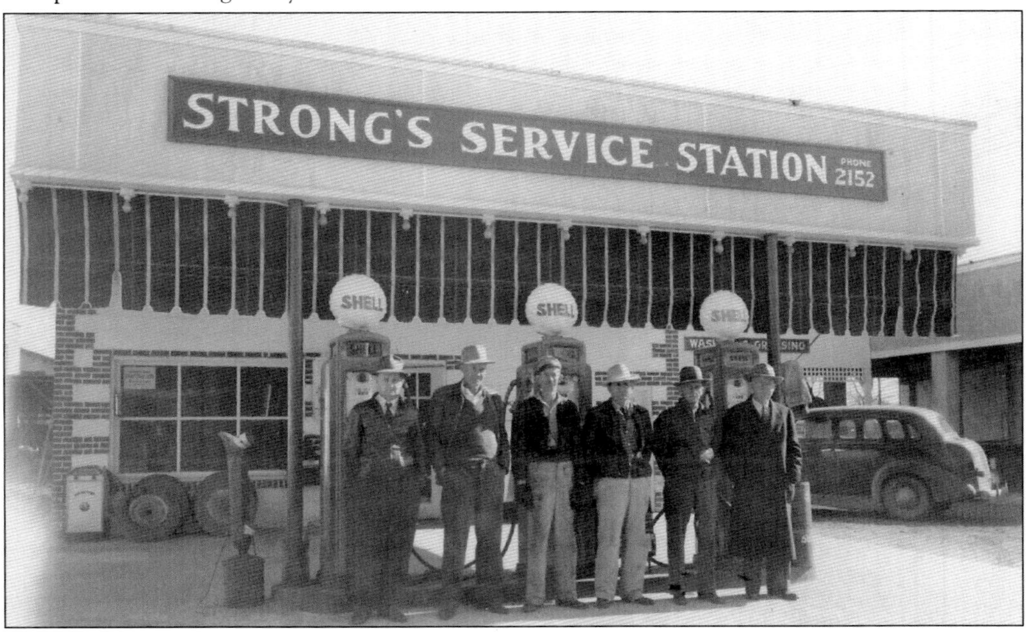

Strong's Service Station stood in what is now the parking lot between Faucett's Department Store and Sue's Flowers. Proprietor Fred Strong started his first service station in Northport in 1932. Strong was one of the first to provide retail coal distribution for residents and businesses in the Northport area. He acquired mined coal from the Fleetwood area and then sold it retail from his service station. He also sold tires, auto parts, kerosene, cordwood, and fish bait and supplies.

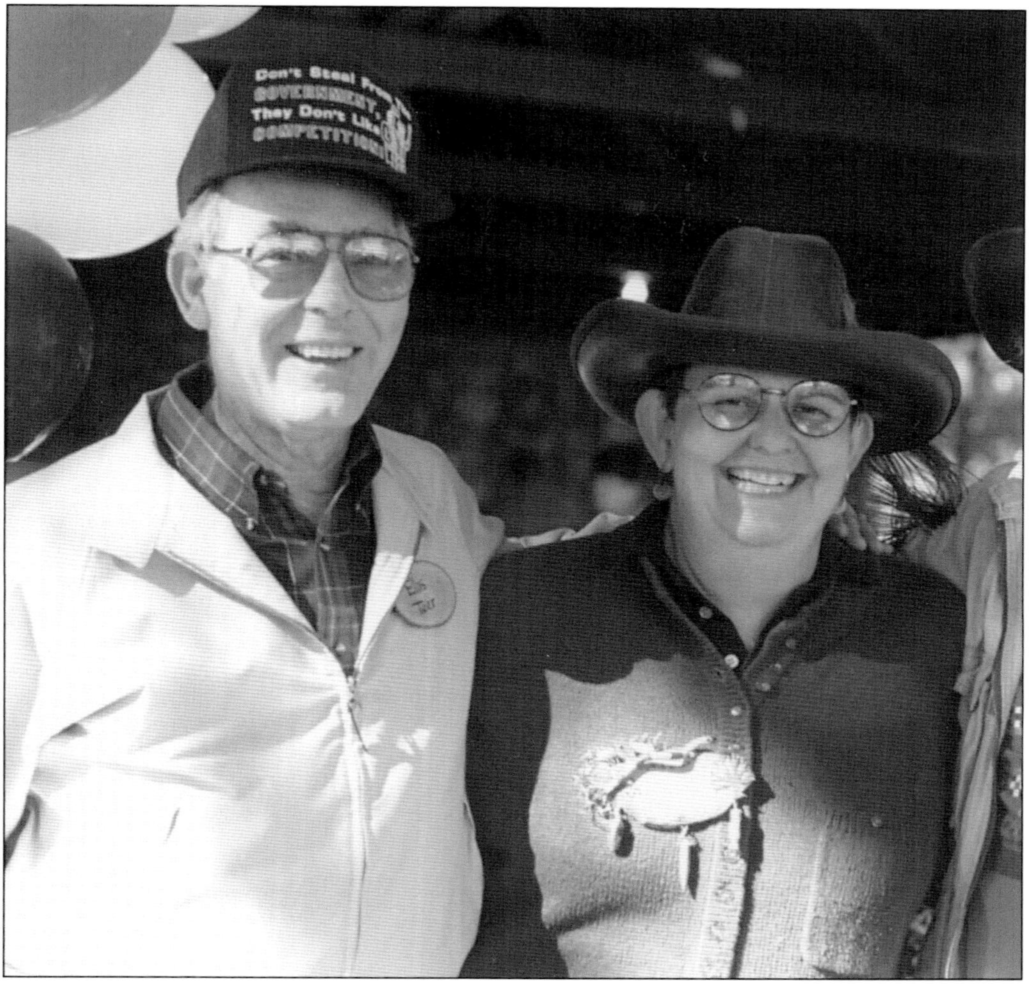

Much of the success of the Kentuck Festival of the Arts, held annually at Kentuck Park in Northport, is attributable to Ellis Teer and Georgine Clarke. A Kentuck founder, Teer has served as president of the Kentuck board and performed numerous other duties for the organization. Clarke was named chairman of the Kentuck Festival in 1973 and became its first executive director in 1978. Born out of Northport's 1971 Centennial, the Kentuck Association has become one of the area's leading cultural organizations, and the Kentuck Festival of the Arts was named one of the top 10 art festivals in the nation by *American Style* magazine in 2012. (Courtesy of Kentuck Association.)

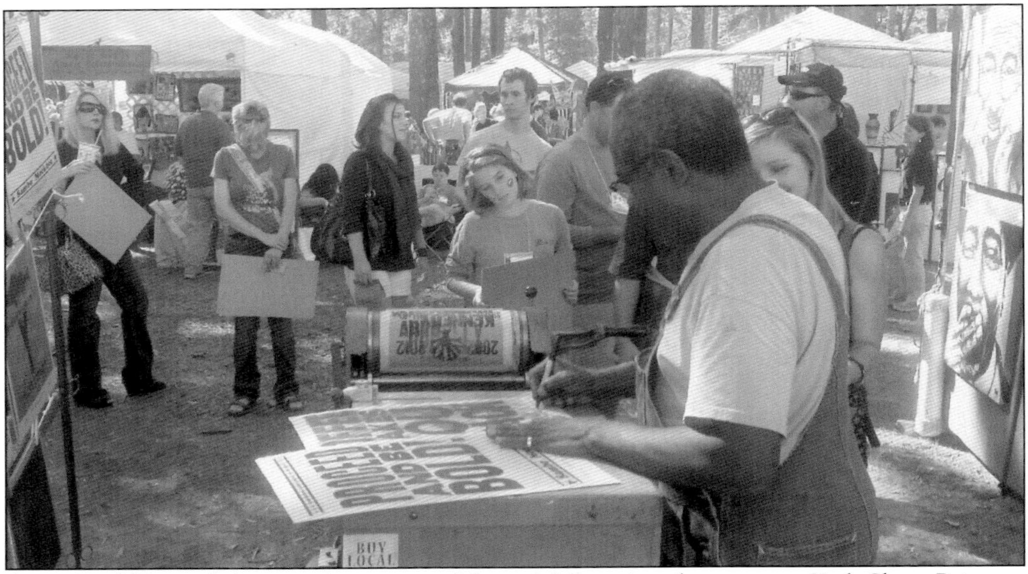

Amos Paul Kennedy Jr., seen here, is a celebrated folk artist who operates York Show Prints, a letterpress poster shop in Gordo, Alabama. Kennedy, a former computer programmer, preserves African American culture through the medium of his art. For more than 30 years, various artists have exhibited poster art at the Kentuck Festival of the Arts. (Courtesy of Dr. Marcy L. Koontz.)

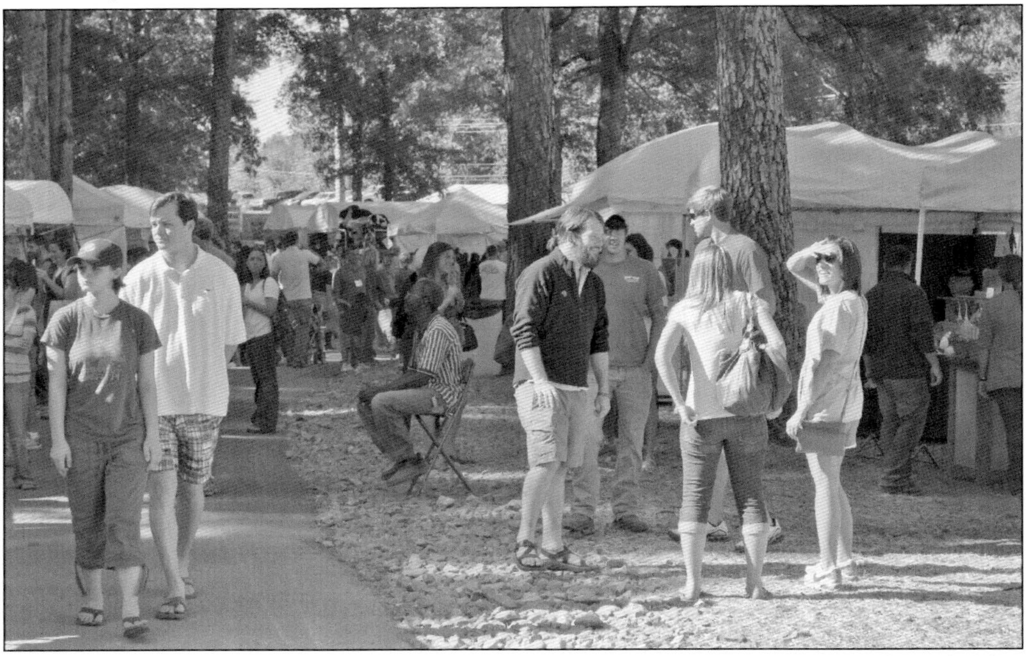

Attendees walk through the 2010 Kentuck Festival of the Arts. The festival has been held at Kentuck Park each October since 1972. The genesis of the festival's name came from a planning meeting in 1972, when Carl Adams suggested the name "Kentucky" because early Northport was sometimes referred to as Kentucky or Kentuck. The name Kentuck was accepted, and later, the Northport City Council was asked to name the park Kentuck, which it did. (Courtesy of Dr. Marcy L. Koontz.)

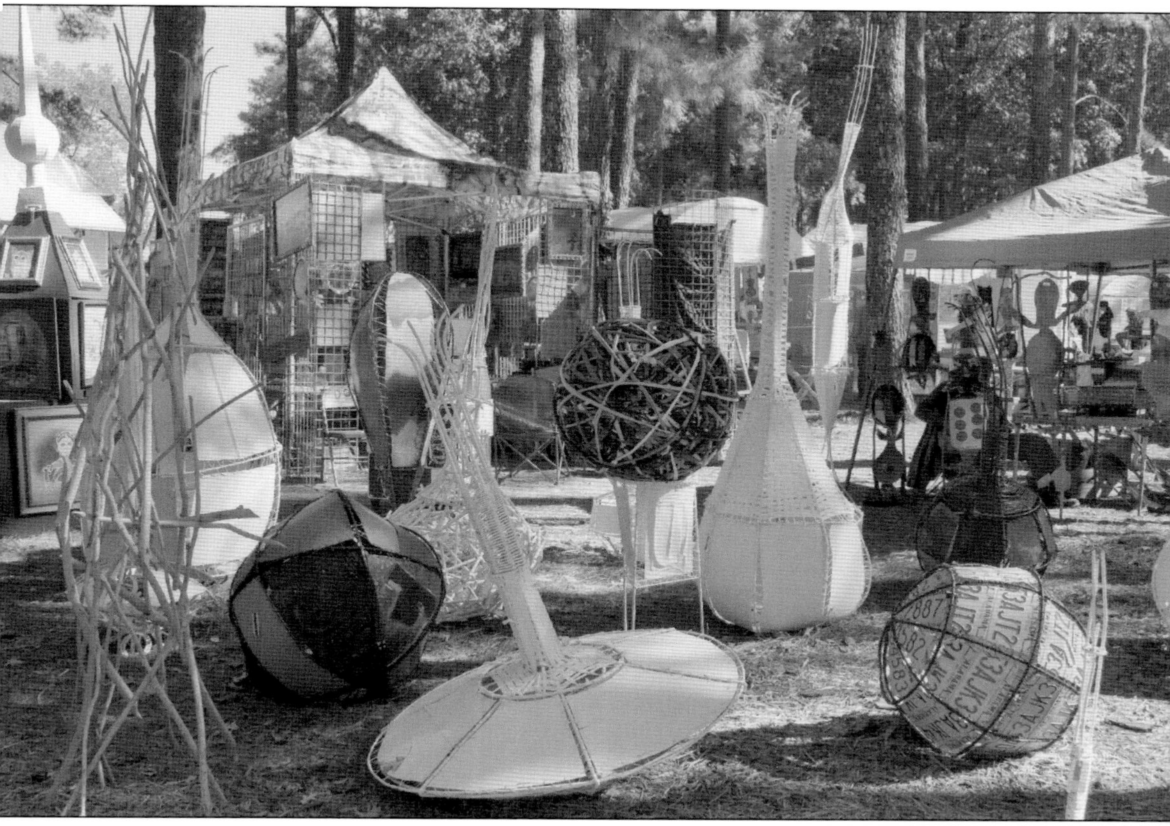

Excellent folk art continues to be displayed at the Kentuck Festival of the Arts every fall. The first festival featured 20 artists, and only two years later, there were 35. Today, more than 200 authentic folk artists display their traditional crafts and attract buyers and admirers of art from around the nation. Every third weekend in October, what was once a small-town event is now known by many from across the nation as "Kentuck Weekend." (Courtesy of Dr. Marcy L. Koontz.)

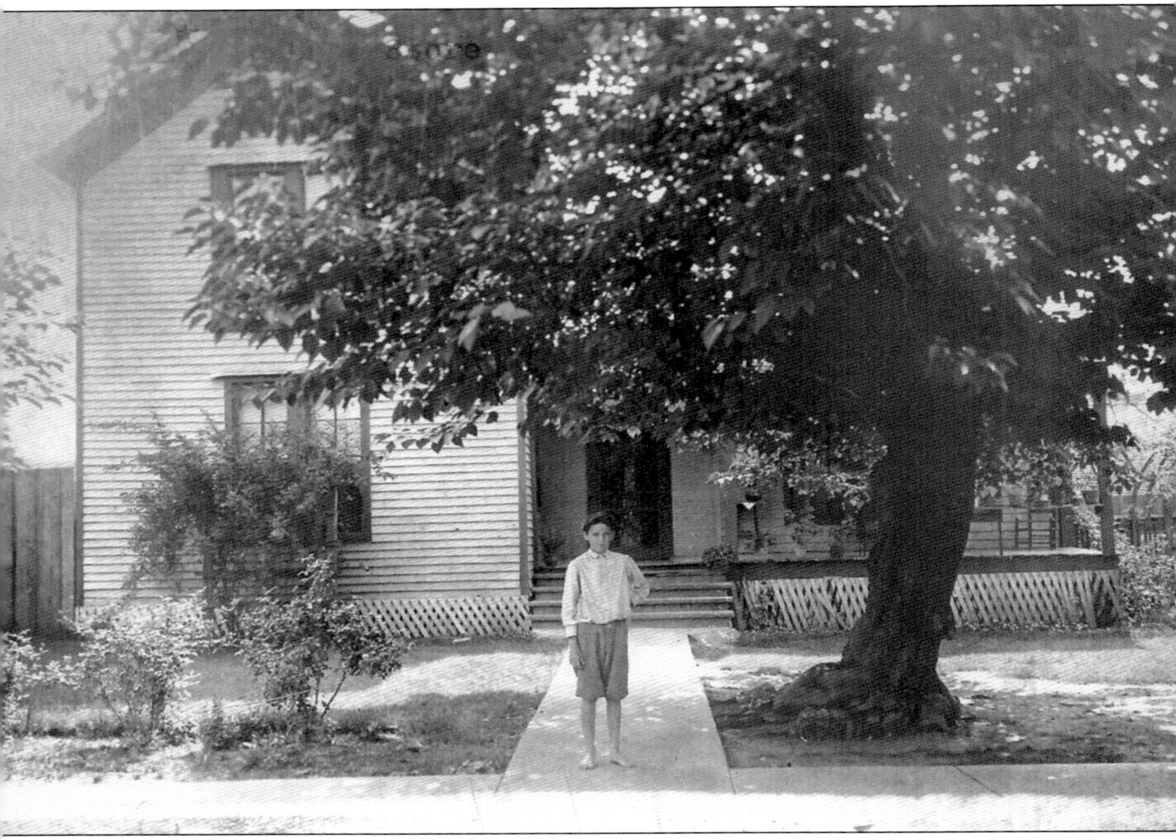

A young Fred Strong stands in front of the Strong home, next to Faucett's on Main Avenue. The house was later moved behind Blackmon's Shoe Store in the 1950s. Strong owned and operated Strong's Service Station on Main Avenue.

Four

Worship and Learning

Northport schools began with private schools, most notably the Academy School, which began around 1830. The first public school in Northport was built in 1901 at Ninth Street and Main Avenue, and was in operation until 1921, when the Northport School was built. In 1926–1927, that building became the Northport Elementary School when Tuscaloosa County High was built on Twenty-fourth Street. The Northport Training School sat where the Sprayberry Center is today. It was destroyed by the 1932 tornado and rebuilt soon after.

Northport Methodist was established in 1837, followed by Northport Baptist a year later, both created as part of the temperance movement. The churches were leaders in changing the "open saloon condition" that had plagued Northport for many years. In the 1840s, these two churches established a joint effort, the Temperance Society, with members from both churches. The society alternated meetings between the two churches. There was sharp discussion at the meetings about the open saloons and related activities in Northport, as well as efforts to promote a better community.

Northport Methodist Church has been at the same location since 1837. Northport Baptist Church began on a 70-foot-by-70-foot corner lot east of Main Avenue near Twenty-first Street and moved to its current location in the late 1850s. Porter St. Paul Christian Methodist Episcopal Church was organized in 1870 and originally located at the corner of Ninth Street and Main Avenue before moving to Bridge Street in 1917. The First Baptist Church was organized in 1887.

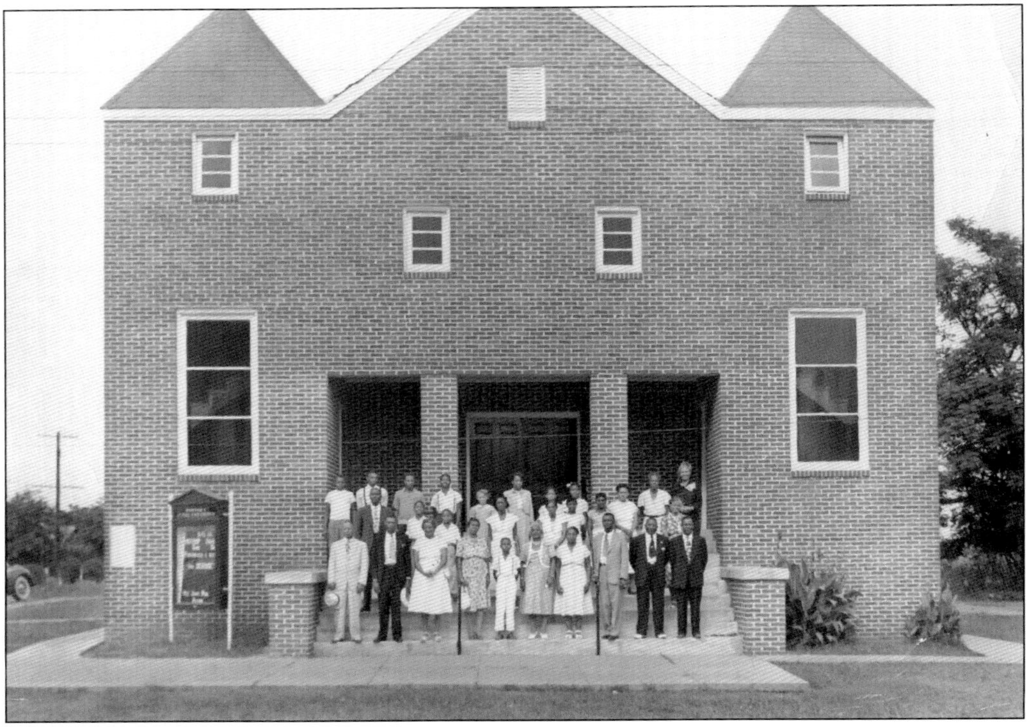

Members of the Brownlee family stand in front of Porter St. Paul C.M.E. Church. The structure was built in 1925 and a brick veneer was added in 1945. In 1964, an educational annex was added.

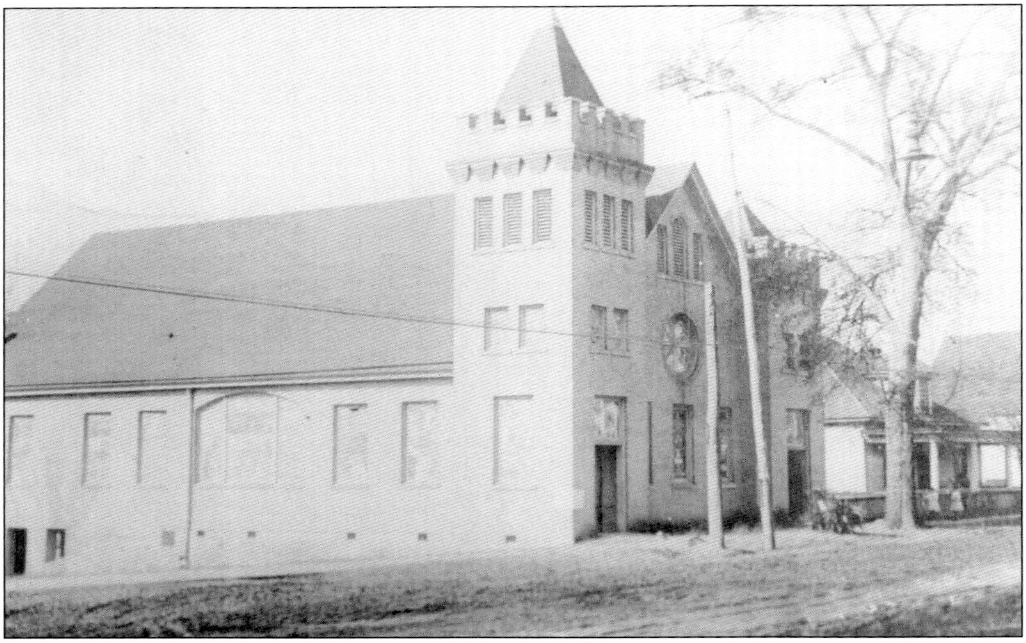

The Northport First United Methodist Church, seen here in 1913, was built by Arthur Laycock. There are known photographs of a previous church building; this is one of the earliest photographs of the structure, which continues to serve as a house of worship for many Northport citizens.

Women from Northport Baptist Church enjoy tea in the church's fellowship hall in the 1960s. They are, from left to right, (first row) Miriam Darnell, Betty Barnett, and Betty Booth; (second row) Johnnie McGee and Charlene McGee. Darnell was the church pianist for many years.

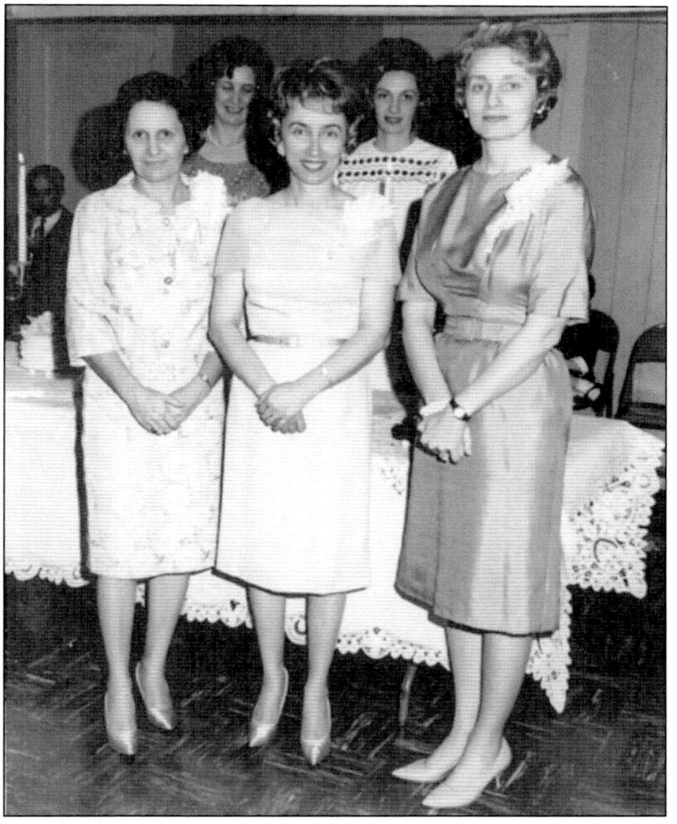

Jennings Chapel United Methodist Church is seen here in the 1950s. This wood-framed structure served as the house of worship for many years. When it was replaced, the building was given to another church in Northport for continued use.

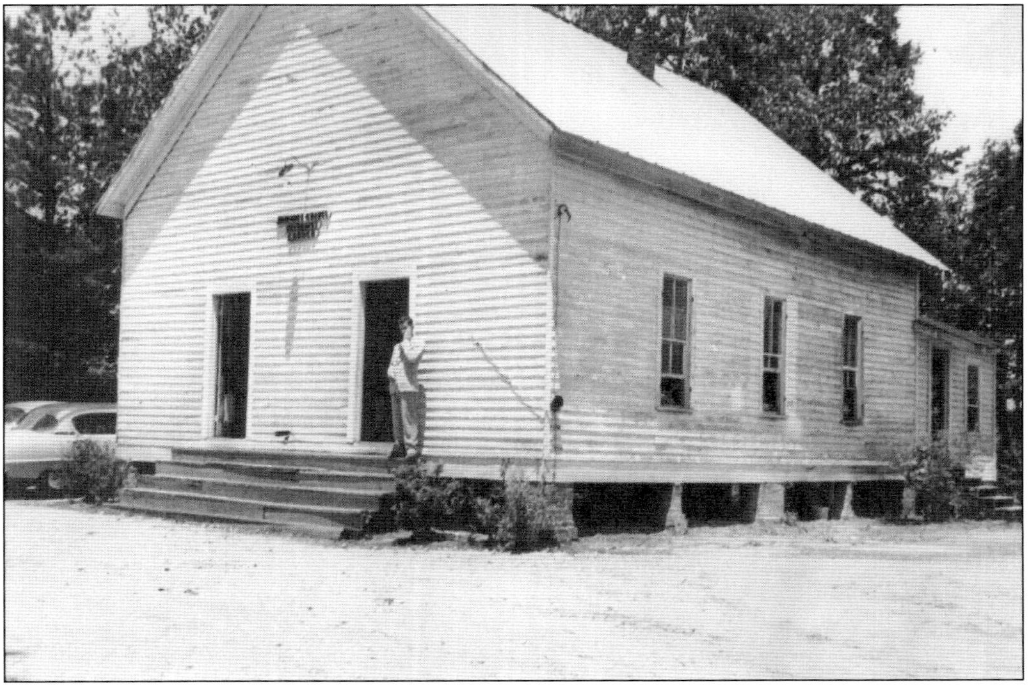

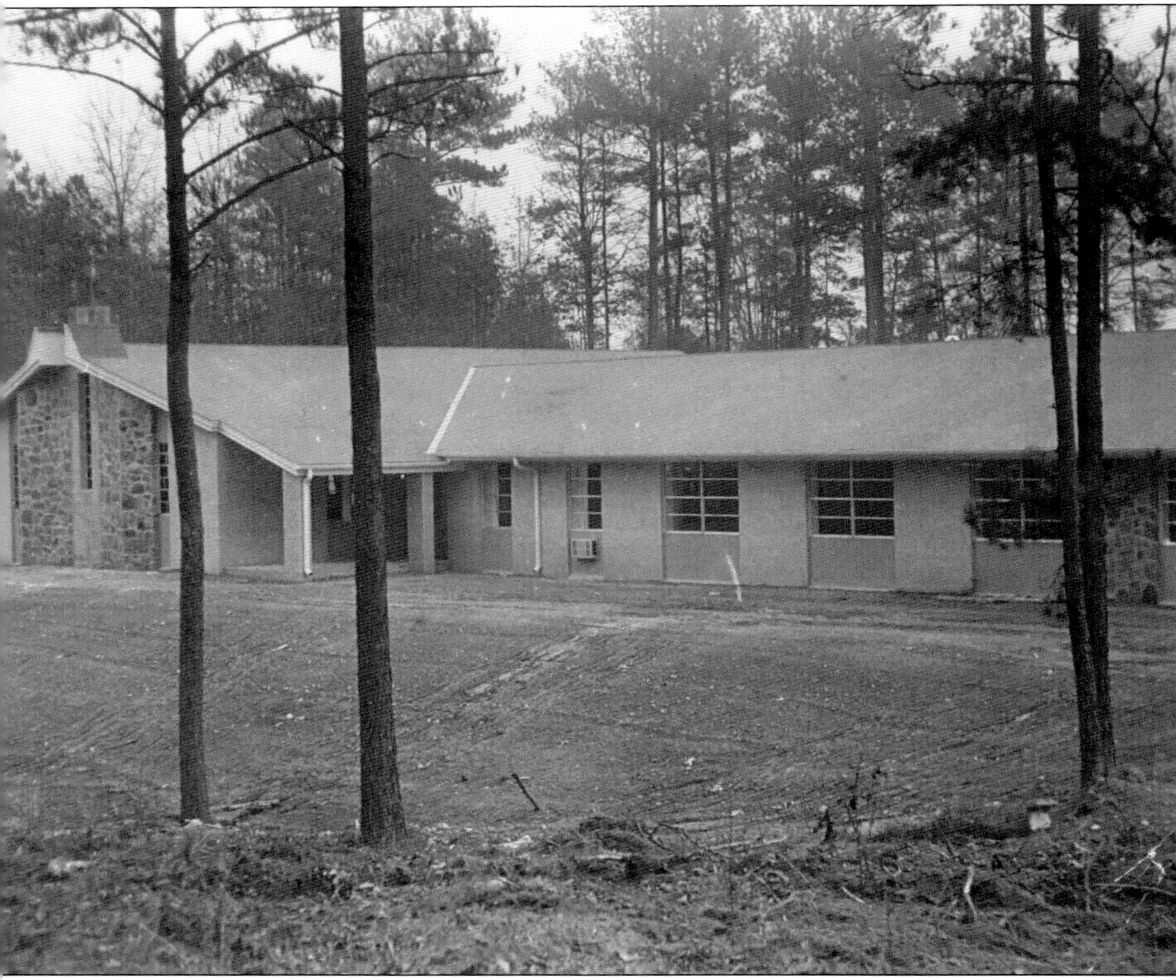

The first house of worship built for St. Mark United Methodist Church is seen here right after its completion in 1966. The church was organized in June 1965 and ground was broken for the new structure in May 1966. Located on McFarland Boulevard, this original house of worship is still in use. A new sanctuary was completed in 1980.

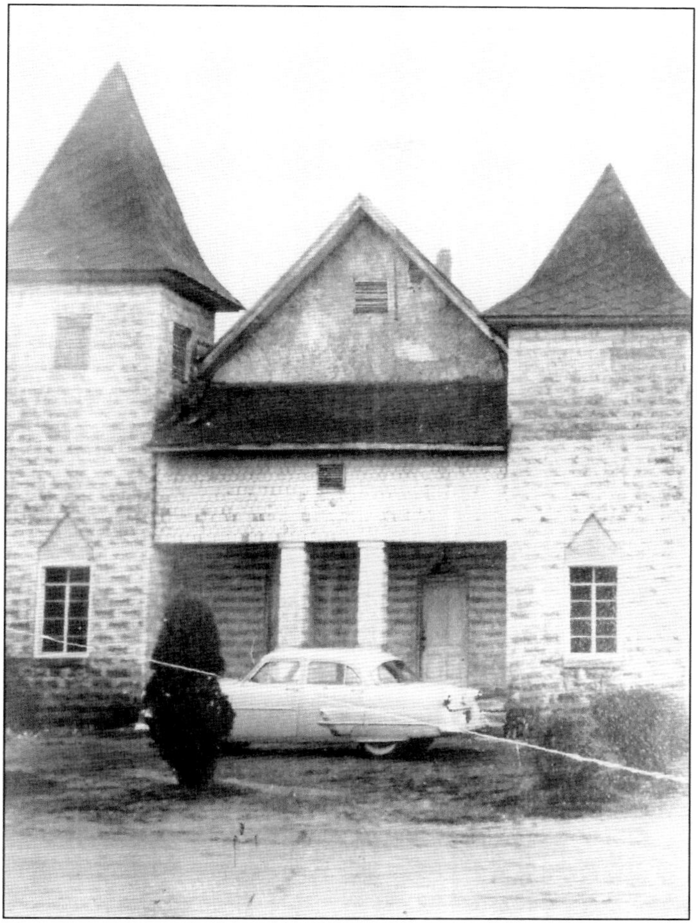

The First Baptist Church of Northport is located at 1401 Twelfth Street. The church was organized in 1887 and the current structure was erected in 1913. The vehicle parked in front of the church belonged to Rev. Osborn Samuel Harvey, pastor from 1937 to 1977. Below, the men of the First Baptist Church of Northport pose in the church in the 1950s. In 1957, an education building, dining room, choir room, pastor's study, and kitchen were added to the church.

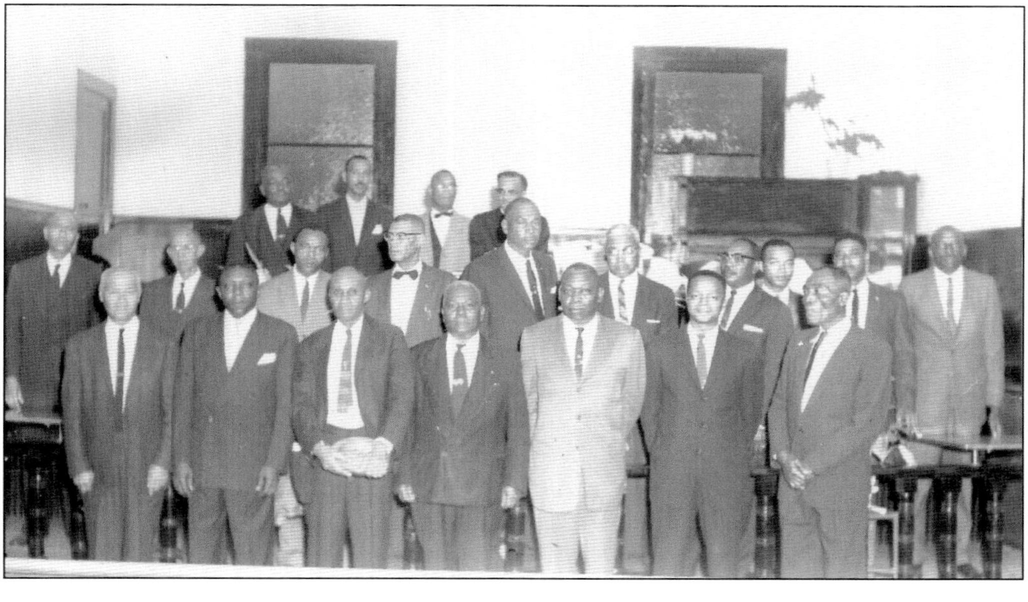

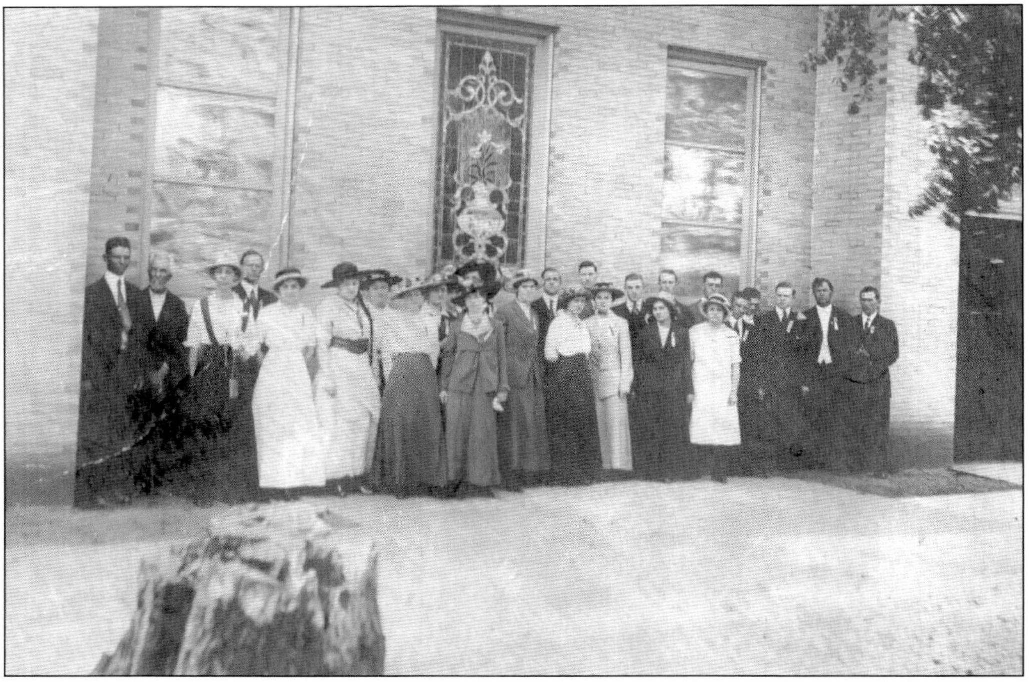

Members of Northport First United Methodist Church pose in front of their newly built house of worship in 1913. A church has stood at or near this same site since the 1830s.

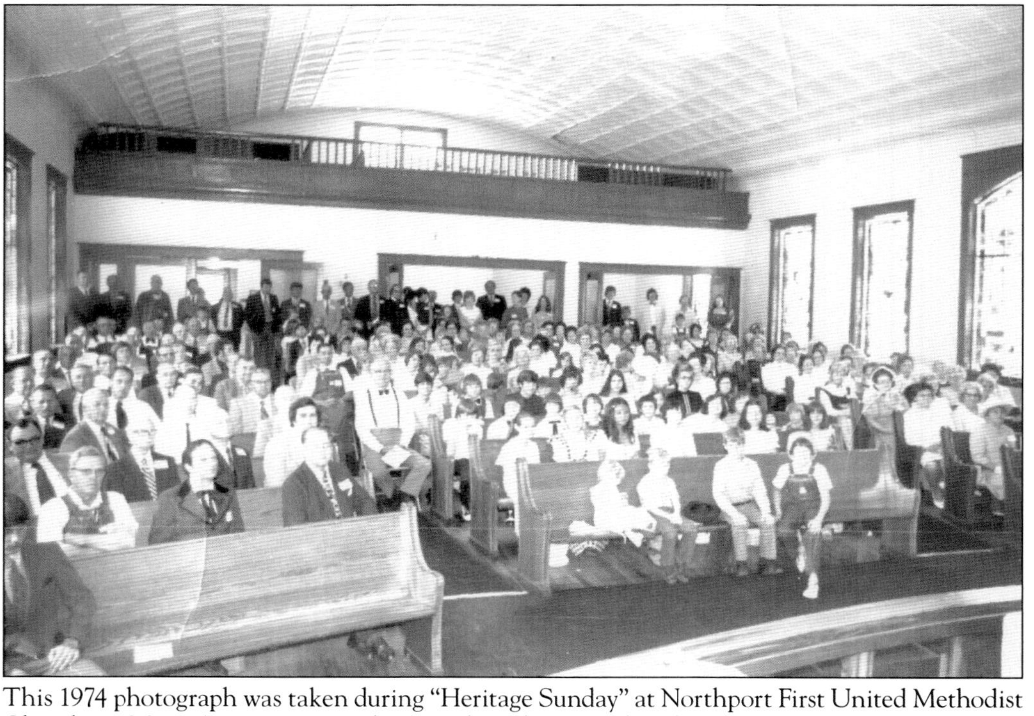

This 1974 photograph was taken during "Heritage Sunday" at Northport First United Methodist Church on Main Avenue. A new historical marker was placed in front of the church that day and many of the church members dressed in period attire.

Posing on the front steps of Northport Baptist Church in 1924, this Sunday school class includes, from left to right, (first row) Wilhelmina Quarles, Maureen Palmer Thompson, and Wilda Anders Mize; (second row) Marie Purdue, Alice Cork, Mary Albright, and unidentified; (third row) Ruth Traywick, teacher Mary Curry, and Pauline Traywick. (Photograph donated and identified by Wilhelmina Quarles Echols.)

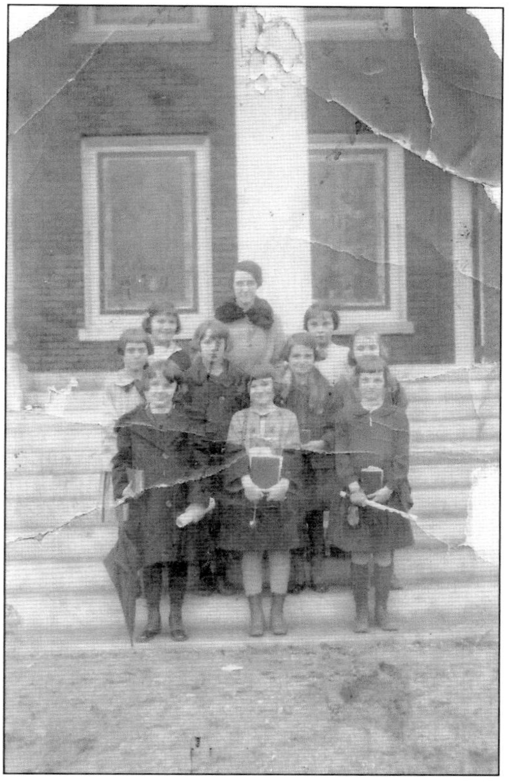

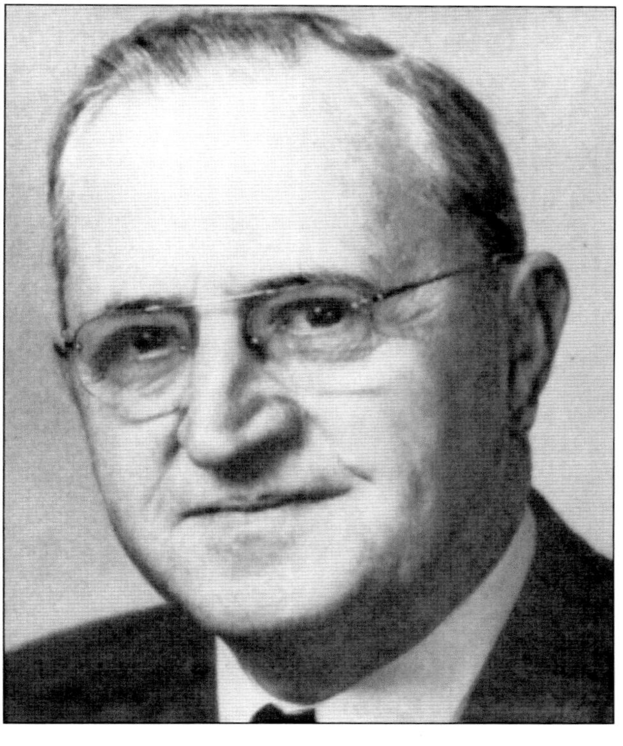

Rev. B.F. Atkins served as pastor of Northport Baptist Church from 1928 to 1958. He also led in the establishment of the Five Points and Northwood Hills Baptist Churches. "Preacher," as his many friends affectionately called him, served in associational work and preached in churches throughout Alabama. He served the Northport community in many capacities and received the first Northport Citizen of the Year award in 1973.

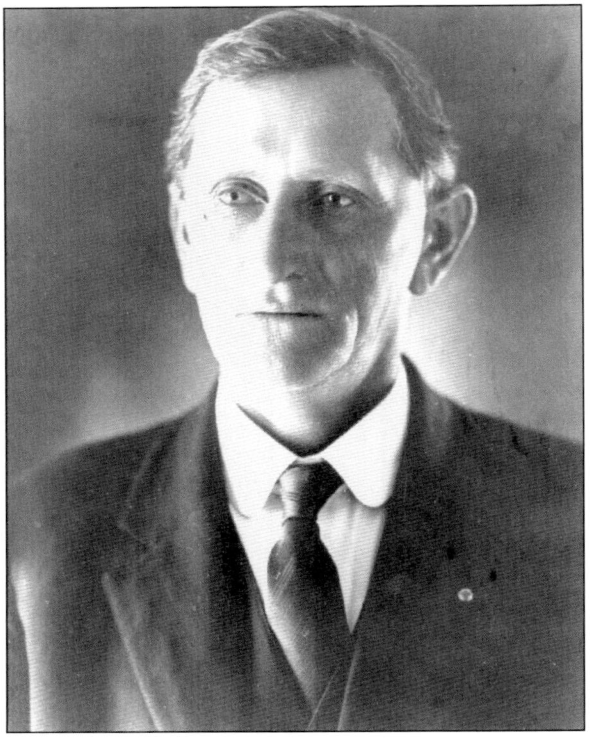

Rev. John Thomas Bealle was ordained by the Northport Baptist Church in July 1891 and served pastoral charges for 23 years in numerous localities, including at Flatwoods Baptist Church in Northport. Reverend Bealle was superintendent of education of Tuscaloosa County from 1896 to 1904 and also served as a representative in the state legislature.

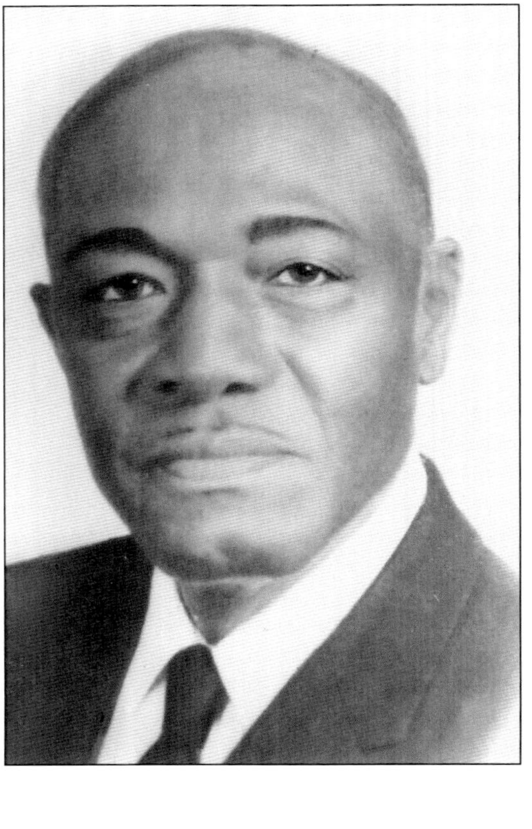

Rev. Osborn Samuel "O.S." Harvey was the pastor of the First Baptist Church in Northport from 1937 until his death in 1977. He was an avid spokesman for the black community in Northport and provided strong leadership during the civil rights movement of the 1960s. Reverend Harvey studied theology at Stillman Institute (now Stillman College) in the 1920s and received an honorary doctor of divinity degree from Selma University in 1951.

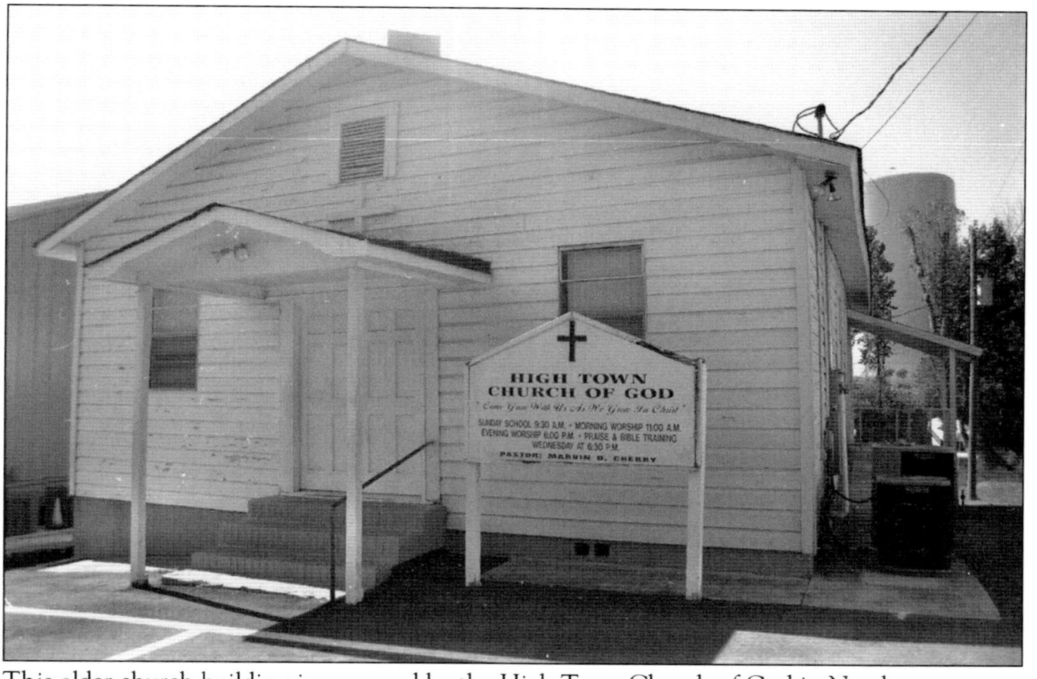

This older church building is now used by the High Town Church of God in Northport.

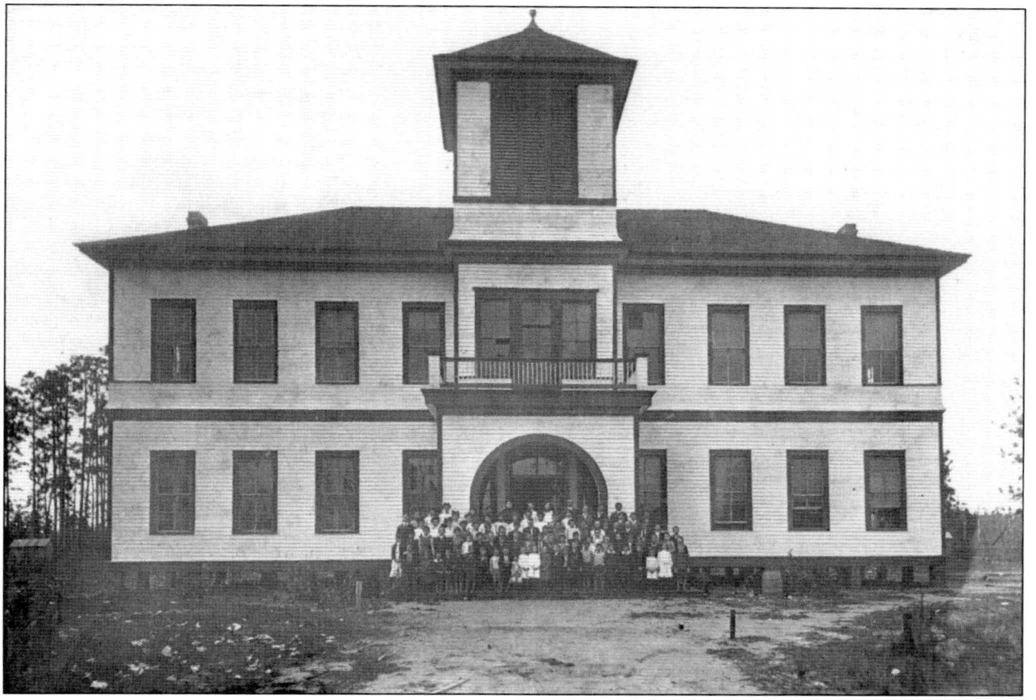
Believed to be one of the earliest schools in Northport, the Northport Academy, seen here in 1895, was located on Academy Hill across from what is now the Northport Community Center on Park Street. The Academy is believed to have been built around 1838, and Vina McGee was one of the first teachers.

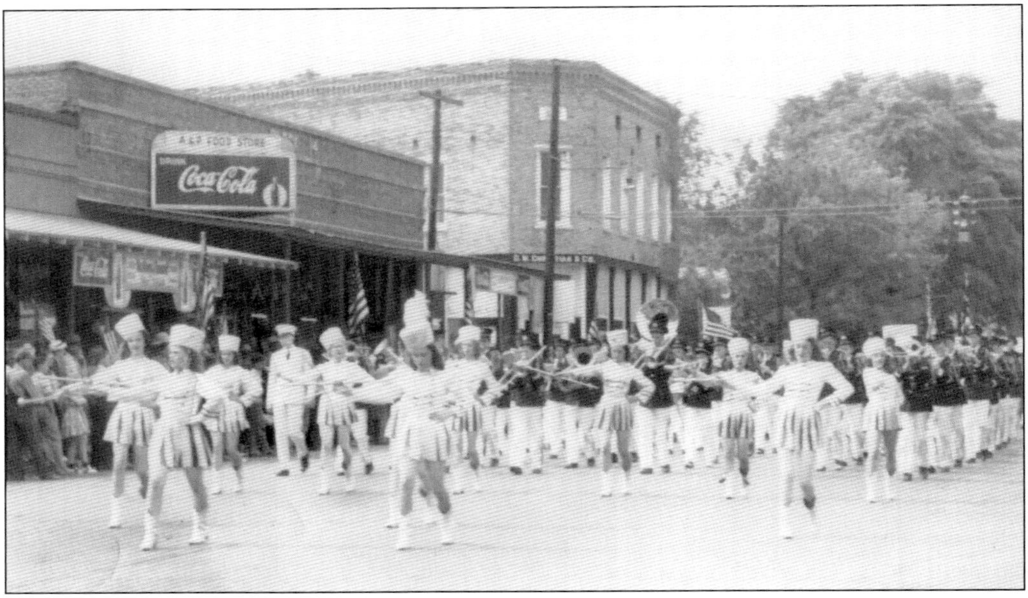

The Tuscaloosa County High School band marches down Main Avenue in 1942. The band director at the time was Stanley Willson, and Jeane Benefield was the head majorette. Other majorettes were Nellene Hardin, Billie Jean Smalley, Virginia Lancaster, Doris Laycock, Frieda Bonner, Corendene Tierce, Faye Smith, Fredalene Hamner, and Denise Faucett. (Courtesy of Billie Jean Smalley Lefebvre and the Northport Heritage Museum.)

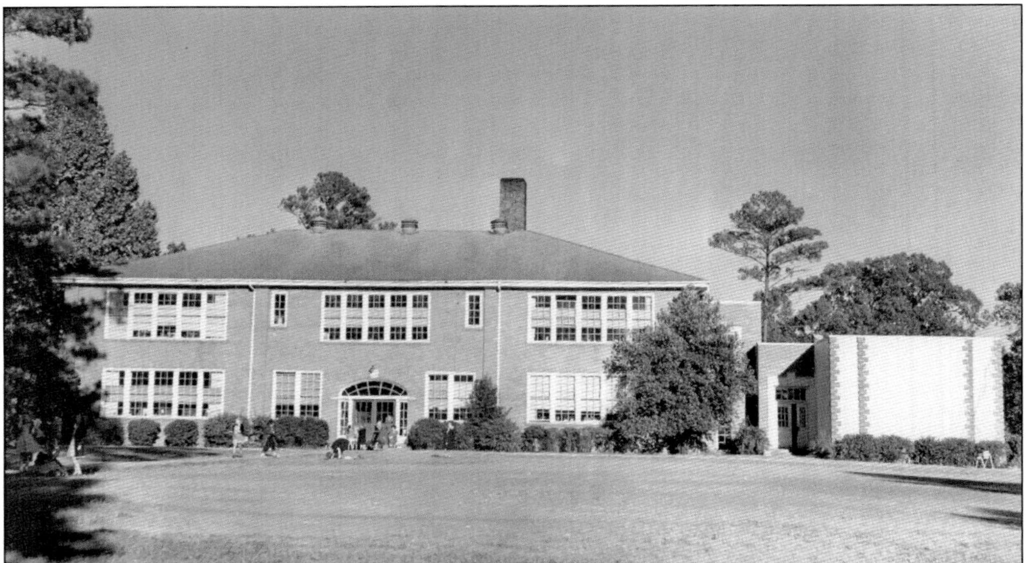

In 1921, Northport High School opened on Main Avenue, replacing the public school that had operated on Ninth Street since 1901. Dr. Houston Cole was the first principal of the new high school and, five years later, became the principal of Tuscaloosa County High School when it was built a few blocks away in 1926. In February 1927, the old school became Northport Elementary School. This school, with its library addition on the right, closed in 1970 and burned down in 1972. Today, the property serves as Northport Civitan Park.

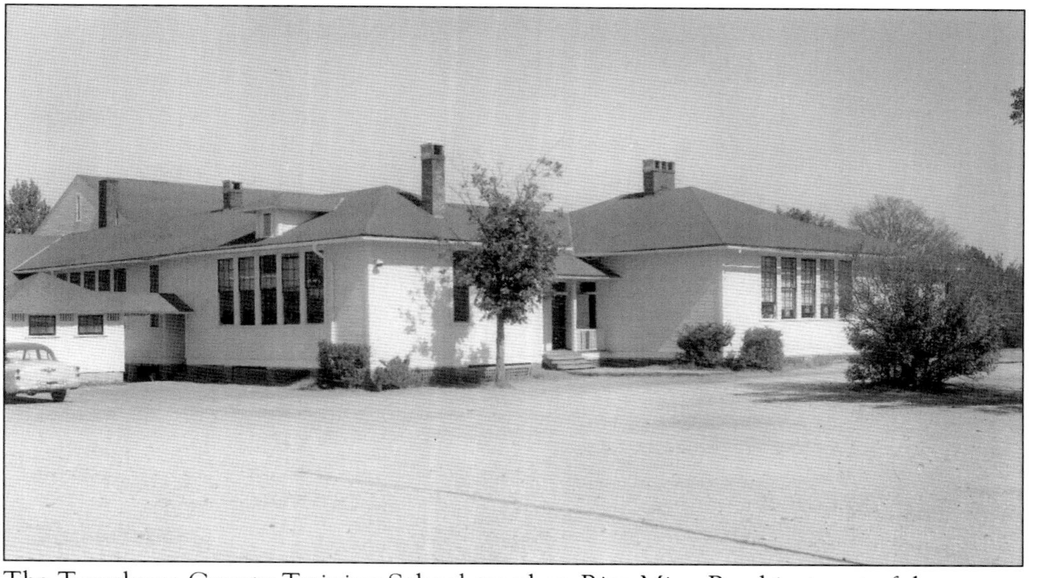

The Tuscaloosa County Training School stood on Rice Mine Road just east of downtown Northport. The March 31, 1932, tornado destroyed the school and killed three of its educators, but the school was rebuilt quickly. It was later replaced by the newly constructed Riverside High School in 1958. As a result of desegregation, the old facility closed permanently in the early 1970s. Portions of the building were used to create the Sprayberry Regional Education Center just a few years later.

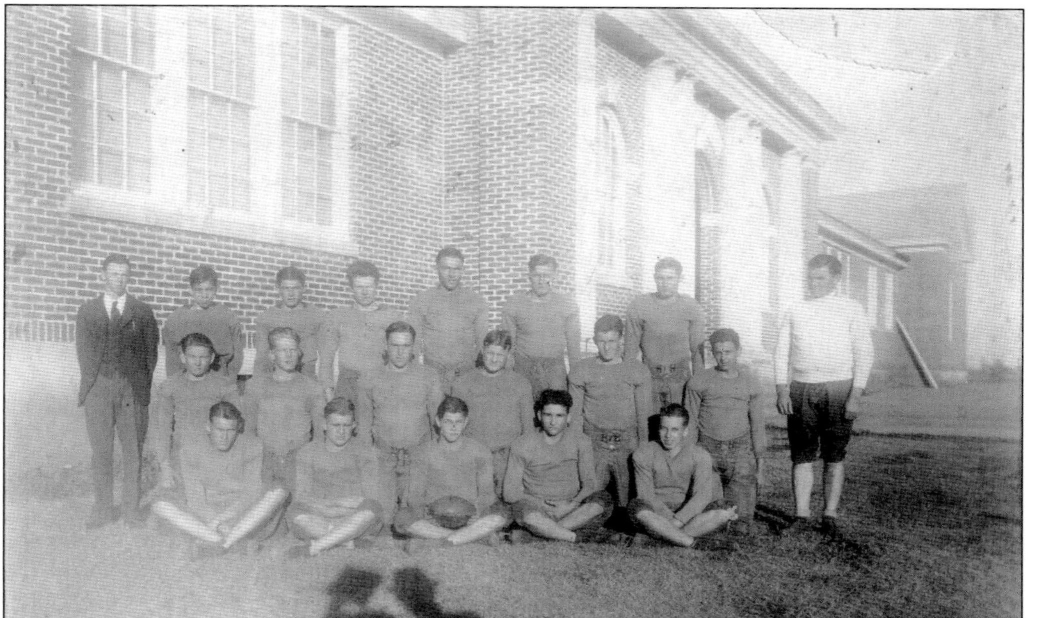

The Tuscaloosa County High School football team poses in front of the school in the late 1920s. Note the shadow of the photographer and his camera equipment in the foreground. This image is dated by the absence of shrubbery along the front of the building, which was planted in the early 1930s.

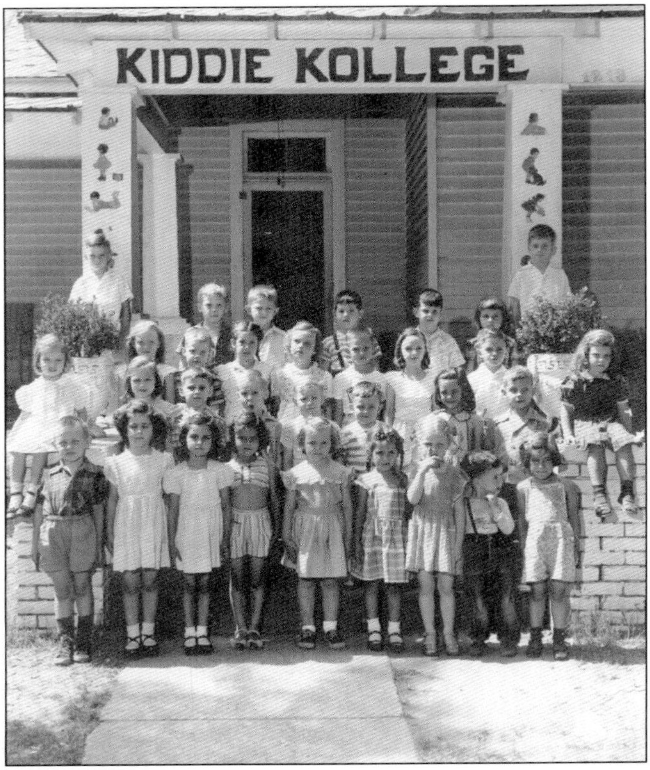

From the early 1930s through the 1950s, Nell Godfrey and Mary Emma "Mama Mac" McKinley operated Kiddie Kollege in a former residence on Tenth Street in Tuscaloosa. McKinley taught kindergarten and Godfrey taught first grade. The school continued to operate for many years after their retirement. Many Northport children attended Kiddie Kollege before entering the public school system.

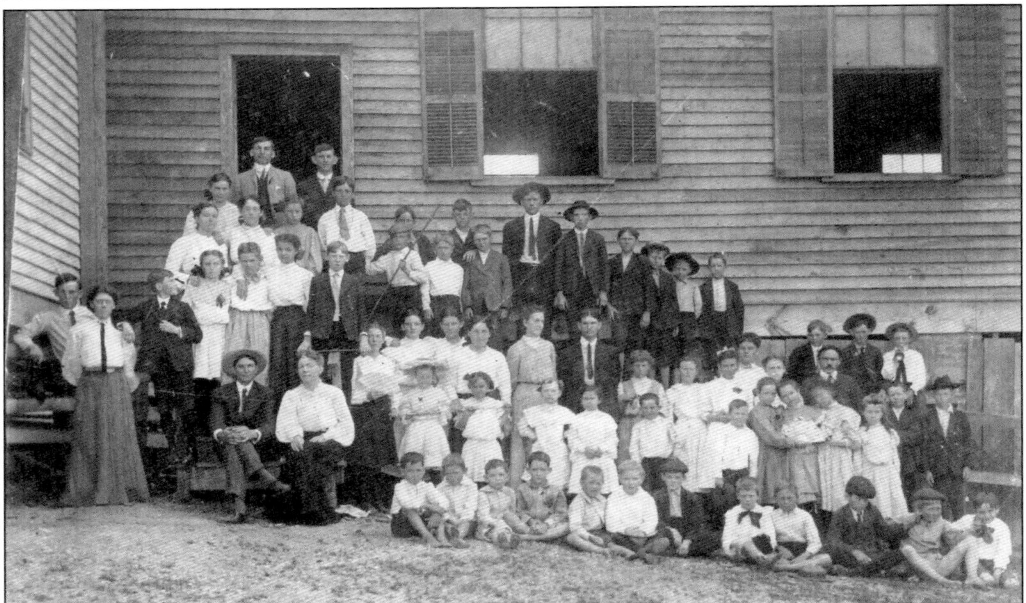

The Northport Public School student body poses in front of the wooden, L-shaped school. This school, off Main Avenue at Ninth Street, thrived from 1901 to 1921. After the school closed in 1921, Glenn Rice converted the building into four different residences, three of which still stand and are still in use as residences.

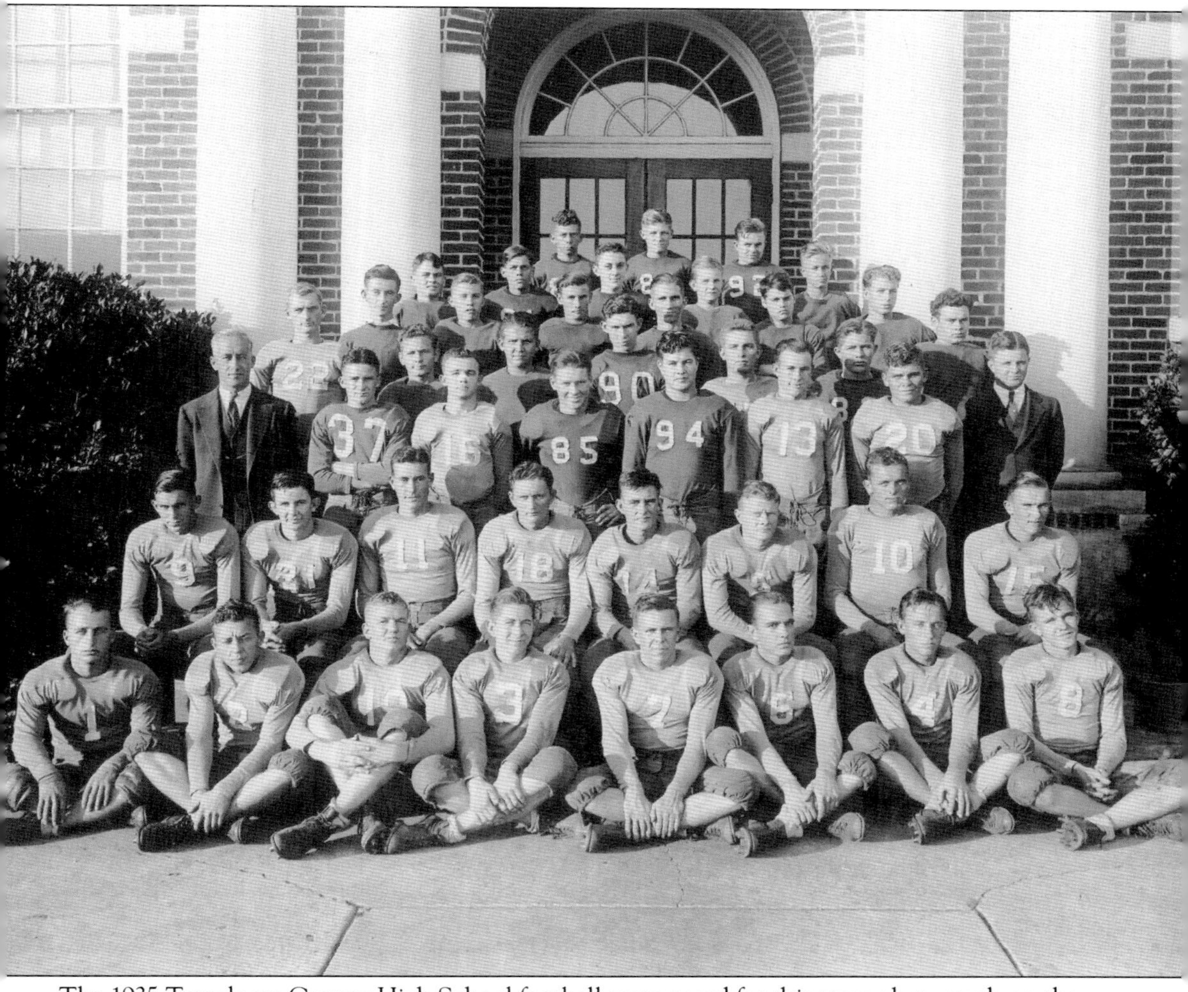

The 1935 Tuscaloosa County High School football team posed for this team photograph on the steps of the school. Coach Frank "Swede" Kendall is on the far left of the third row, and principal Rayburn Fisher is on the right end of the same row. Kendall was the assistant coach in 1935 and the head coach from 1936 to 1938. In September 1939, he returned to his alma mater, Tuscaloosa High School, to coach the Black Bears.

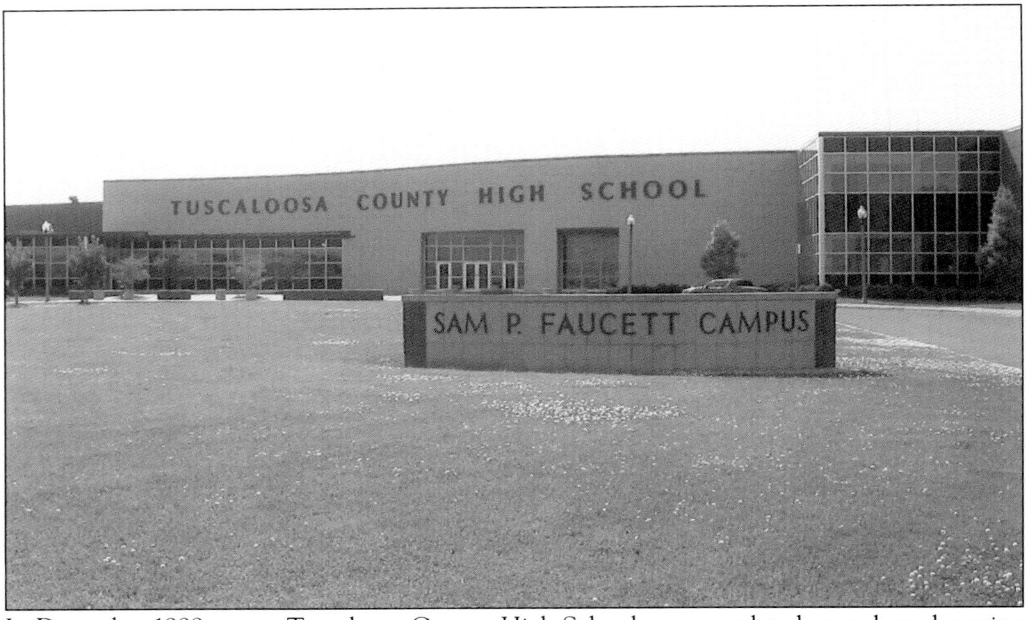

In December 1999, a new Tuscaloosa County High School was completed to replace the aging 1926 school. The new campus was named for Sam P. Faucett III, whose generous gift of $1 million in 1996, along with many other public and private donations, ensured that the much-needed high school would be built and would serve the community for many years to come.

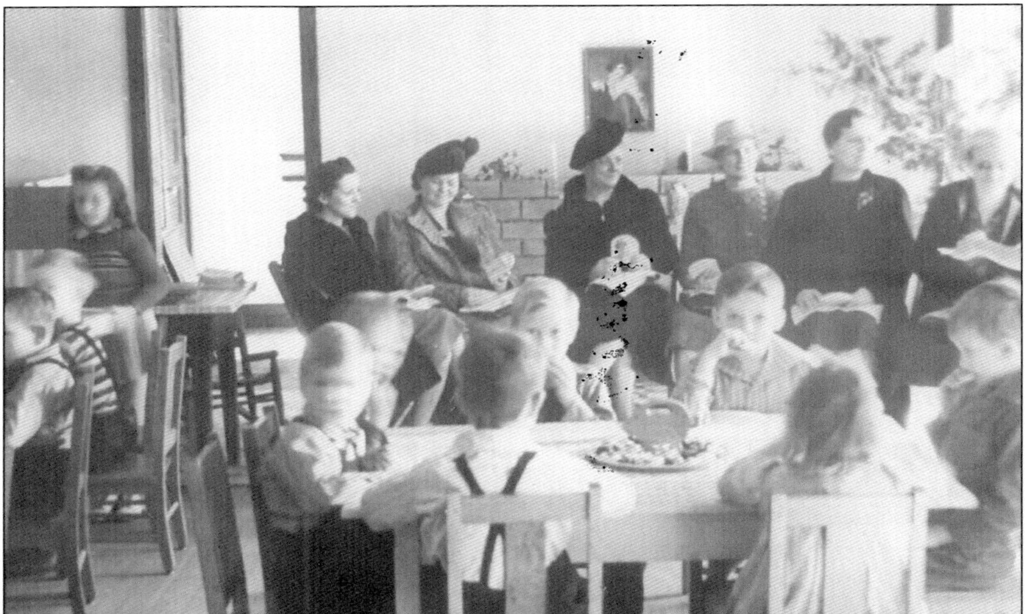

Students and adults eat lunch in the old Northport Elementary School. The school's principals have been Dr. Houston Cole, Nellie Reynolds, Bess Hamner, Dr. Charles Sprayberry, Dr. Neil Hyche, Kenneth Wann, Samuel Key, Roger Ballard, and Eileen Glass. Jewel Craft taught second grade at the school from 1927 to 1967.

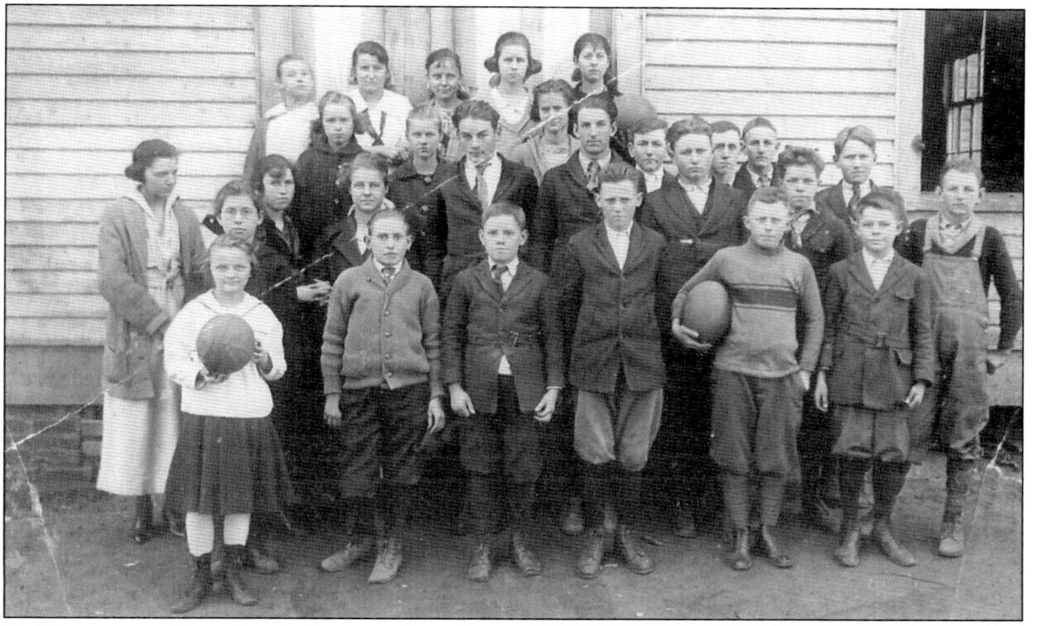

Unlike many older school photographs, almost all of the children in this 1914 photograph at Northport Public School have been identified. They are, from left to right, (first row) Jane Arabella Herring, Herman Perdue, Walter Maxwell, William Howard Darden, Hobson Brown, and Russell Lewis; (second row) teacher Miss Brown; Sarah McDaniel, unidentified, Maidie Gray, Ormand Stephens, Paul Dutton, Joe T. Harper, Ellis Turner, and Eugene Dutton; (third row) Jo Cantrell, unidentified, Pauline Christian, unidentified, Clarence Sullivan, unidentified, and George Rose; (fourth row) Ethel Hasson, Frances Clements, Reubel Strong, Ruth Powell, and Ollie Mae Chism.

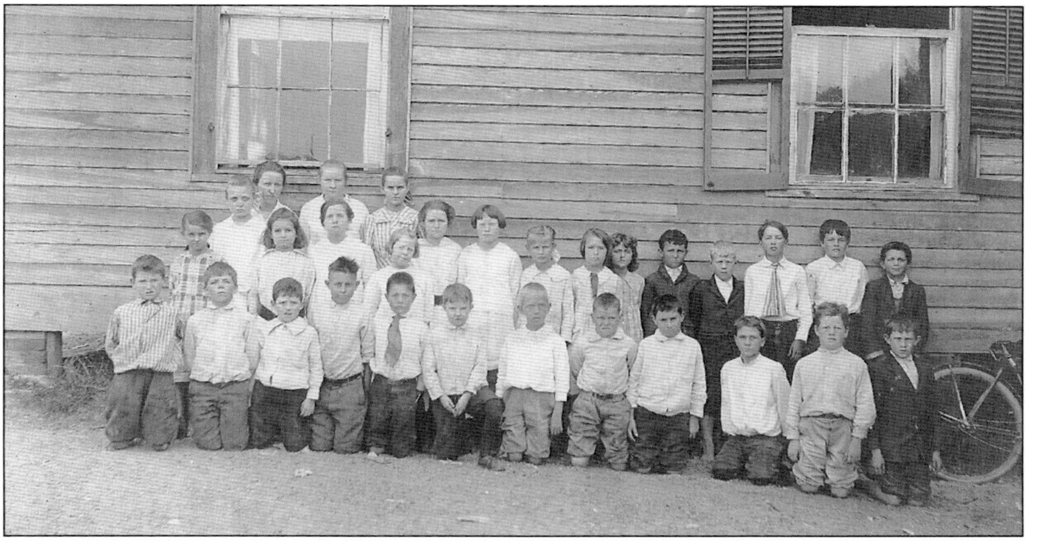

This class poses outside of Northport Public School about 1915. The school was beginning to fall into disrepair, as seen by the condition of the shutters and boards. In an October 1916 article, the *Tuscaloosa News* reported, "Registration at the Northport Public School has reached an all-time high mark of 185 pupils, according to E.M. Meadows, principal."

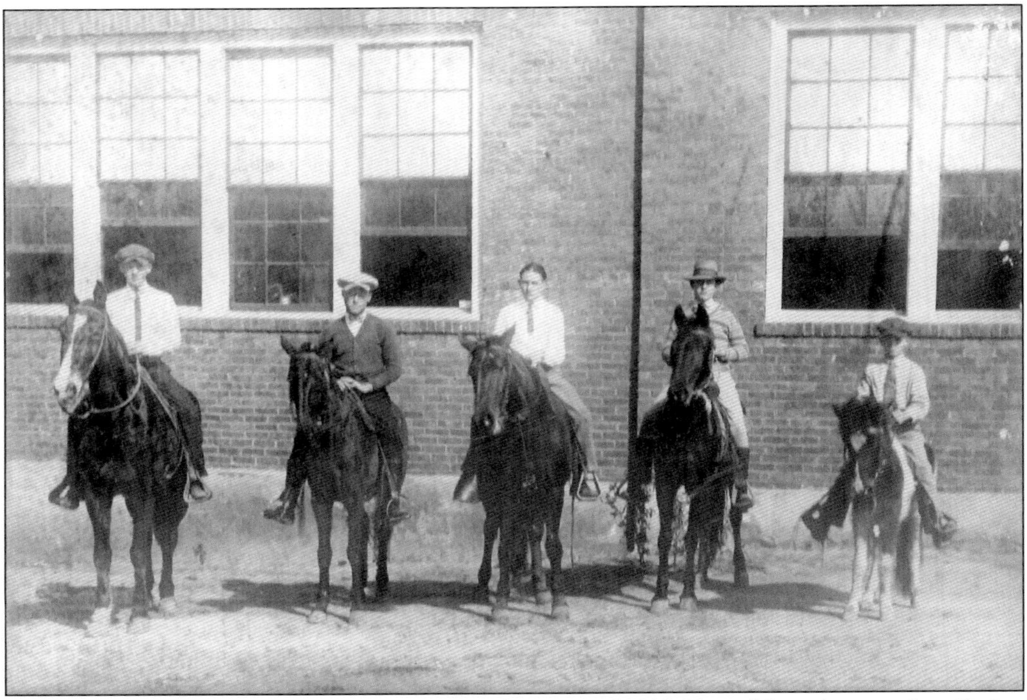

This Northport High School photograph was taken around 1923. Two of the riders are identified as William Howard Darden (second from left) and Lewis Faucett (third from left). While the occasion of the photograph is not known, it did require the students to dress in their best attire.

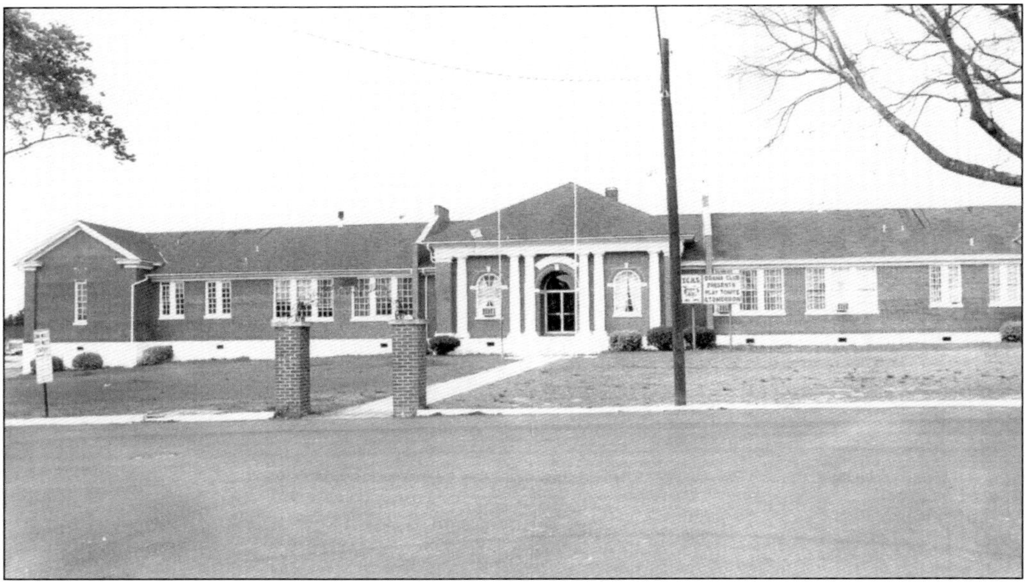

This 1972 photograph of Tuscaloosa County High School shows the World War II Memorial at the front entrance to the school. In the background, a new reader board, given by the 1972 senior class, informs students that the "Drama Club Presents Play Tonite and Tomorrow." At the time, two flagpoles stood at the front of the school. A few years later, lightning damaged the pole on the right, and it was never replaced.

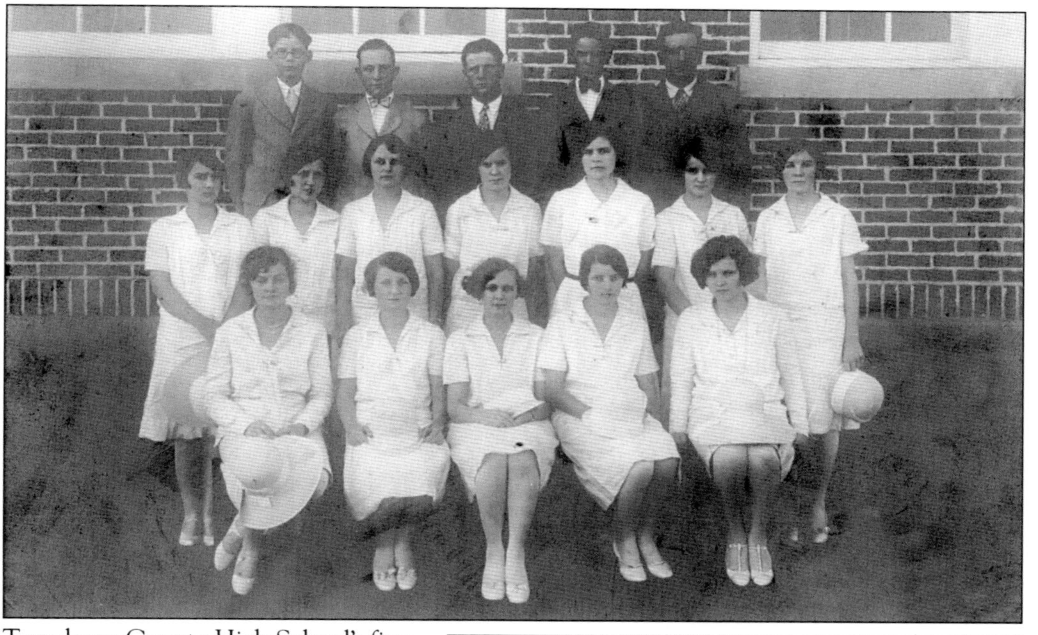

Tuscaloosa County High School's first graduating class, in 1927, included, from left to right, (first row) Hattie Hassell, Leota Shirley, Lenora Deason, Pauline Evans, and Brucie Lee Doty; (second row) Mae Darden, Flora Clements, Vera Bell, Lois Chism, Catherine Hardin, Mary Alice Johnson, and Catherine Thompson; (third row) L.C. Bambarger, Milton Cooper, Newell Shepherd, Chester Utley, and Fred George.

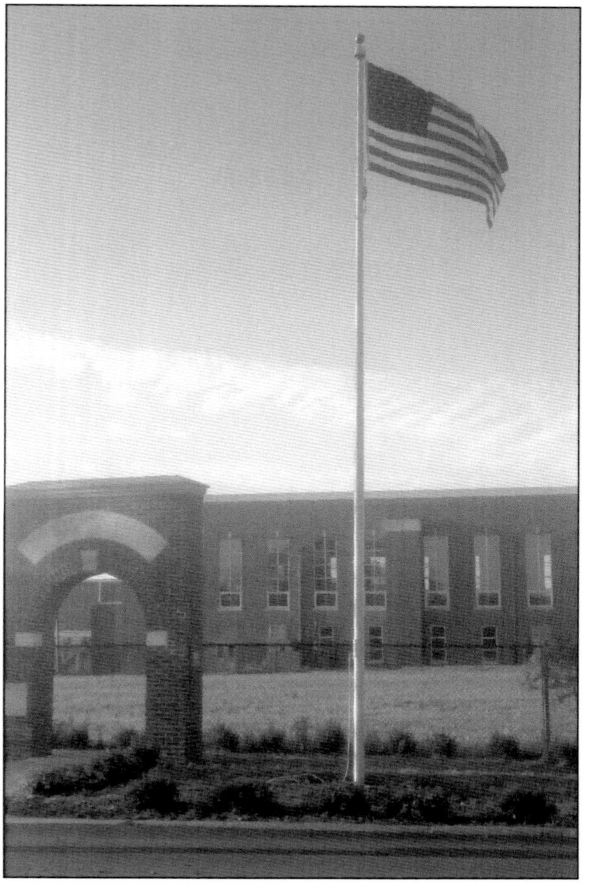

This Tuscaloosa County High School Memorial is located at the old school site on Twenty-Fourth Street. Included on the memorial arch are original bricks from the 1926 school, the marquee stone above the entrance that reads "Tuscaloosa County High School," and the cornerstone of the school. Also part of the memorial is the original keystone—the single white piece at the apex of the brick arch. The small white blocks on either side of the columns are called spring blocks. The memorial was dedicated on May 7, 2011, and the flagpole was dedicated by the class of 1960 on June 9, 2012.

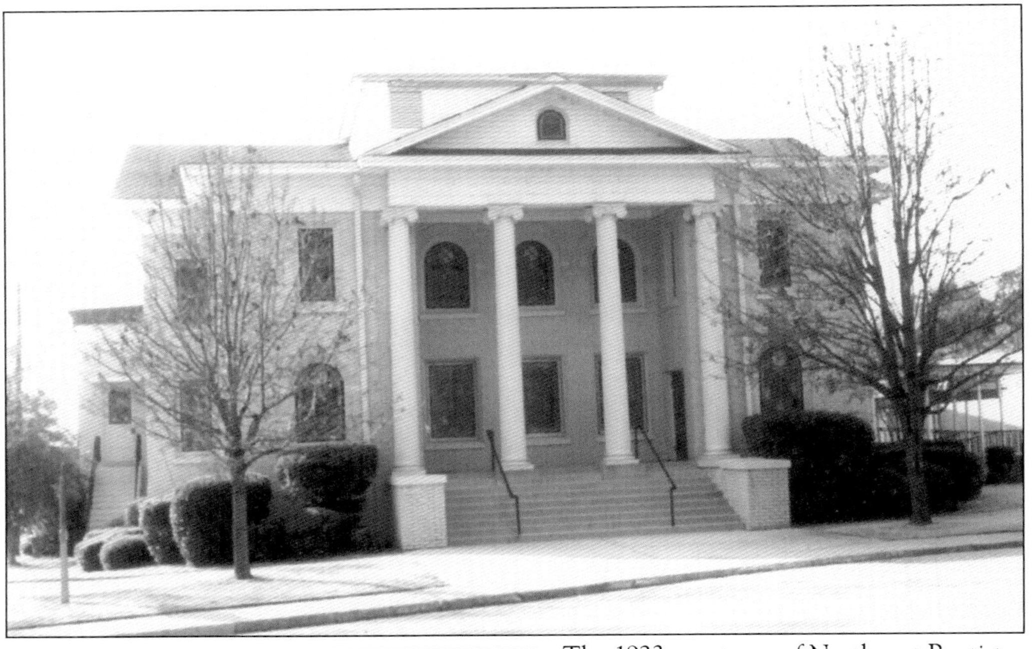

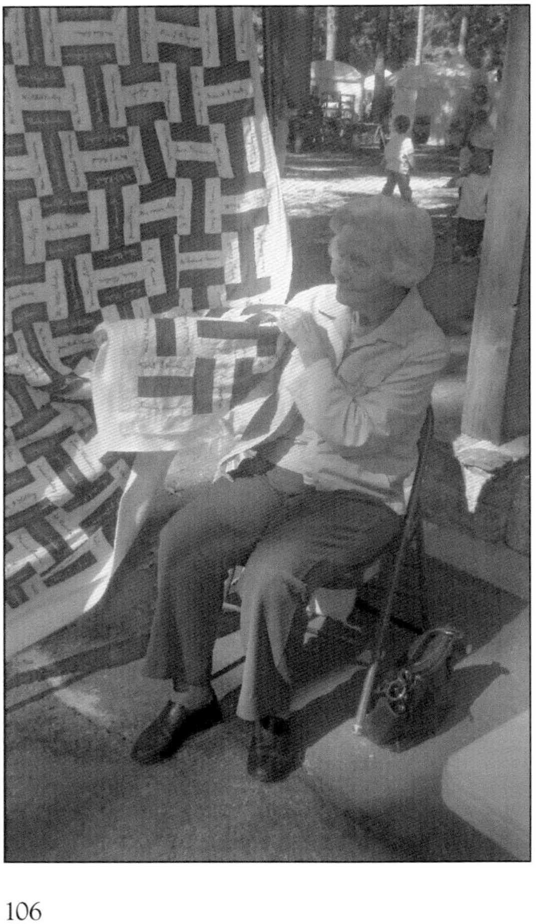

The 1923 sanctuary of Northport Baptist Church was originally a deep red color, but was painted a light beige in later years. In 1946, an Austin pipe organ was installed. By 1974, it was deemed necessary to renovate the inside of the church with new pews and other enhancements. A dedication ceremony was held on October 27, 1974.

This historic signature quilt was recently found by Catherine Robertson Dickson of Gainesville, Florida. Dickson discovered the quilt top in a cedar chest that belonged to her mother, Catherine Burroughs Robertson. Research reveals that the quilt was made as a "memory quilt" to commemorate the building of the 1923 Northport Baptist Church on Main Avenue. During the unveiling of the quilt at the Kentuck Festival of the Arts in October 2011, Sara Merle Spencer Brunette of Northport, seen here, finds the signature of her father, Ford Spencer, in the quilt's lower right corner.

Five

FRIENDS AND FAMILY

Much can be learned from long-time citizens who had a front row seat in observing the growth of Northport. The physical appearance of older neighborhoods, the belief systems and cultural mores that prevailed, and the "little things" that seemed so unimportant back then are among the lessons that can be learned. The best way to know about life "back then" is to sit a spell and talk about it with someone who lived it. Several years ago, Elizabeth Holley Nixon ("Lib" to some) shared with me some memories of her childhood in early 20th century Northport.

On December 28, 1912, Dr. A.A. Kirk, who lived on Columbus Street, went to the Holley home to deliver a baby girl who was named Elizabeth. The Holley family also lived on Columbus Street (Fifth Street today) in a house built on an acre of land that Elizabeth Holley's father purchased from the Shirley family for $150. During those early years, Columbus Street was a dirt road, especially hot and dusty during the summer months. Growing up, Elizabeth and her siblings regularly had to hose down the street to control the dust.

Traweek, Lindsey, Shirley, Koster, Ramsey—these were some of Columbus Street's earliest residents. At one time, there were so many unmarried ladies living on Columbus Street that it was said if a girl lived on Columbus Street long enough, it was sure to make her an "old maid." Elizabeth recalls that her parents taught each of their three children to be neighborly and speak to everyone they passed on the street, especially the elderly people.

Strong family, simple pleasures, few luxuries, hard work, respect, impeccable manners, and character beyond reproach—all deeply important and very descriptive of Elizabeth Holley Nixon, whose growing up years on Columbus Street were imprinted with people, memories, and life-long lessons that she has never forgotten and still values to this day.

—June Lambert

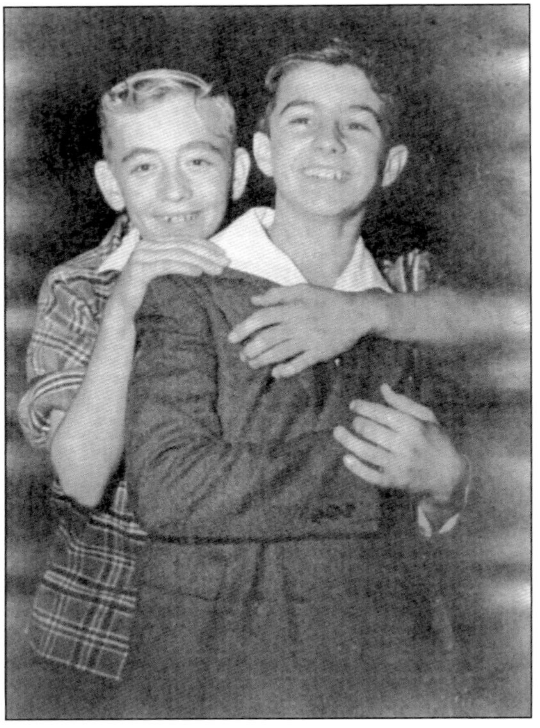

This photograph was taken by Yeatman King at a Tuscaloosa County High School football game around 1940. Lucien Lewis (left) and Bruce Davis became fast friends through their involvement in a Boy Scout troop that met at the First United Methodist Church in Northport. Their scoutmaster was Earnest Mills, and their troop leader was Joseph Shipp. Although Lucien lived in Northport, Bruce, who was a Tuscaloosa boy, became a Northport transplant when he began to date Charlotte Anders, the beautiful girl who later became his wife. Lucien later married the former Sara Frances (Frankie) Bell, a popular girl from a well-known Northport family.

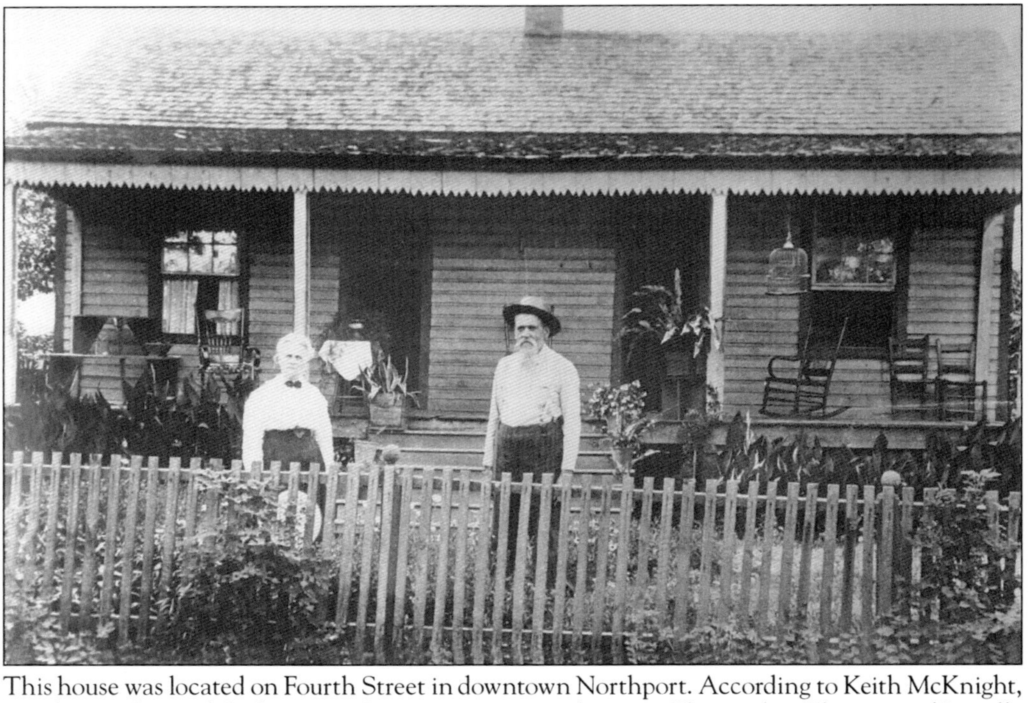

This house was located on Fourth Street in downtown Northport. According to Keith McKnight, standing in front of the house are his great-great-grandparents Christopher Thomas and Louella Shockley Thomas. They owned a talking Mynah bird that greeted people as they walked by the house. Betty Slowe of the *Tuscaloosa News* identified this historic Northport home.

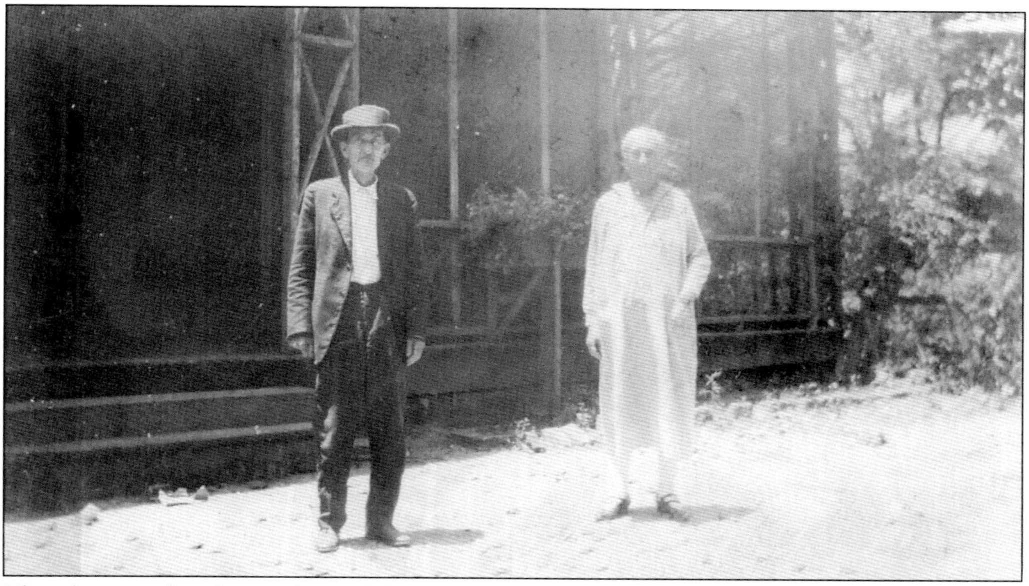

This photograph, taken in front of the Maxwell-Peters home, shows Jacob Holbrook Anders and his wife, Martha Josephine Pumphrey Anders. Their seven children were John Hampton Anders, Burwell Lewis Anders, Levin Pumphrey Anders, Percy Crump Anders, James Williamson Anders, Samuel Faucett Anders, Margaret Josephine Anders, and Susan Brown Anders.

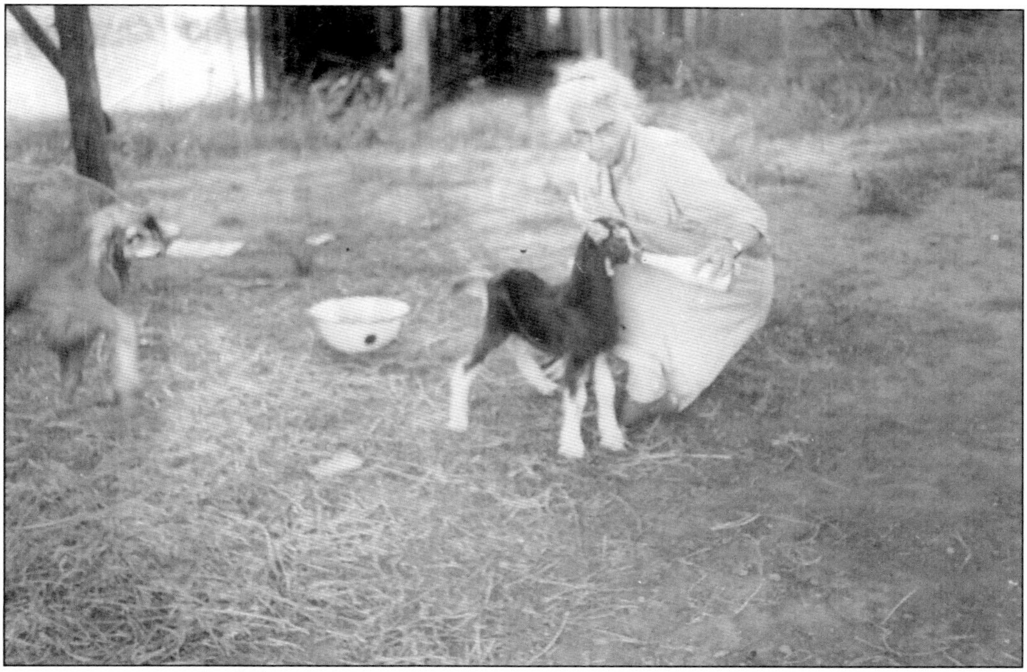

Cora Cummings feeds her baby goat in the yard of her residence on Twentieth Street near downtown Northport in 1949. Many in the town at the time were talking about her beautiful hand-needlework. Cummings also raised and sold beautiful flowers. A "crazy quilt" made by her in 1890 was donated to the Northport Heritage Museum by Barbara Palmer Boyd. (Courtesy of Leatha Darden, gifted to Northport Heritage Museum.)

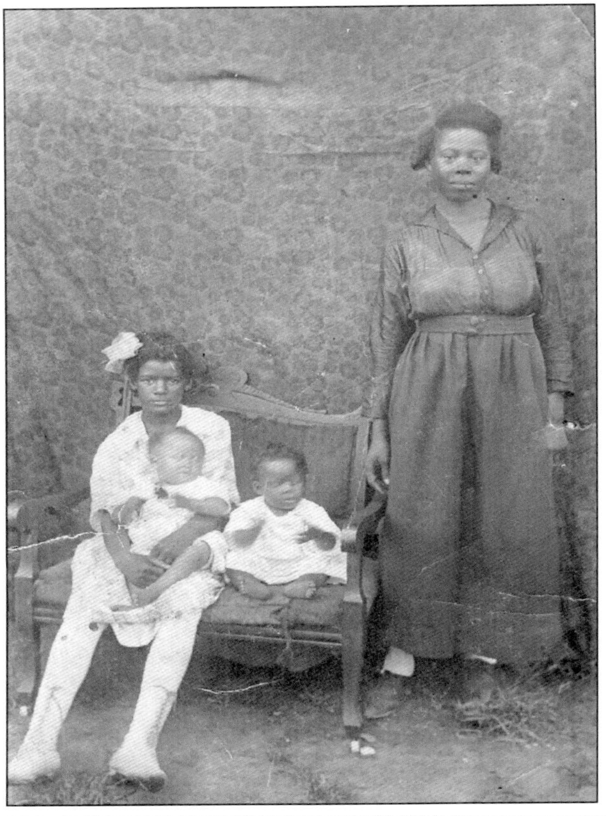

The family of longtime Northport resident Ruby Hamilton Battle is seen here in 1919. From left to right, they are Ella Mae, Ruby, Ruby's twin sister Ruth, and their mother Clara Herring Hamilton. At age nine, Ruby survived a life-threatening injury from the 1932 tornado. After receiving a degree in elementary education, she began a teaching career that lasted for more than 40 years.

Three generations of the Christian family are seen here. From left to right, they are George William Christian II, Tom Christian, and George William Christian Sr.

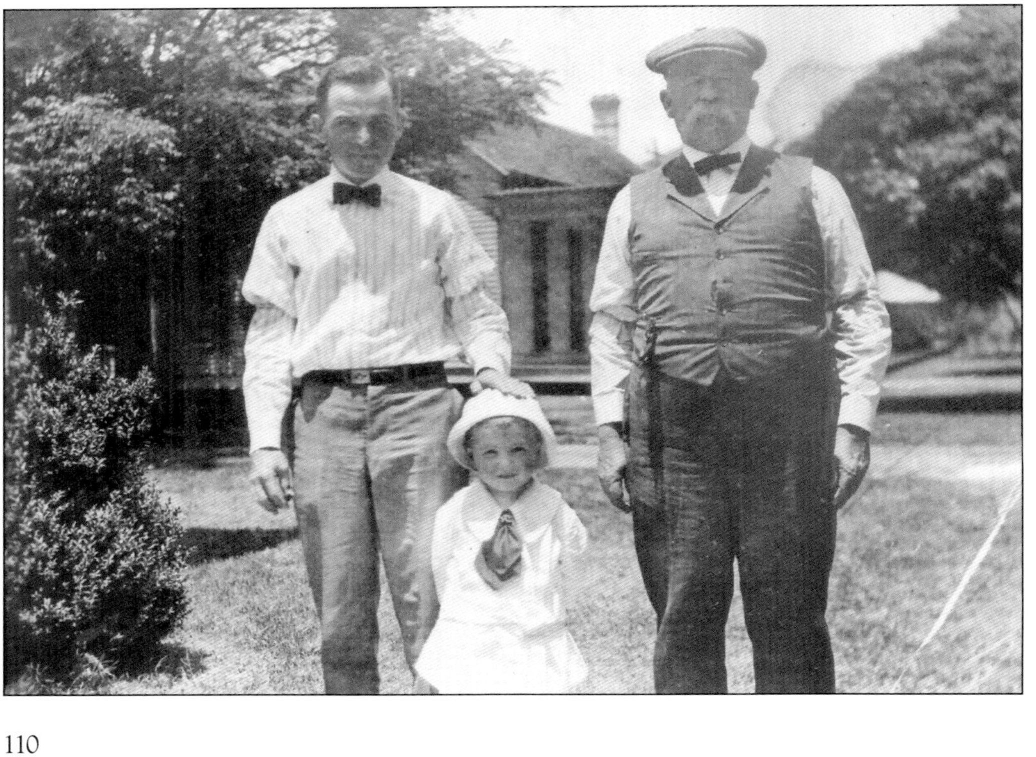

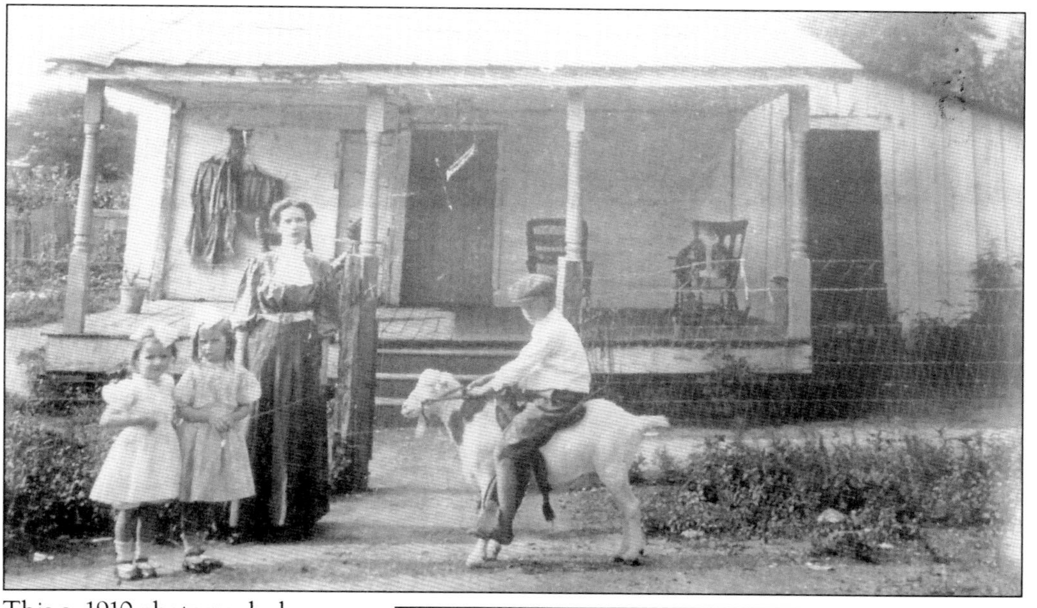

This c. 1910 photograph shows the Elbert Clayton and Clara Comerford Clark family. Sitting on the saddled goat is their son, Francis Elbert Clark. His two younger sisters, Marylee and Loucile, are on the left. The Clarks lived in a modest home in downtown Northport behind City Cafe and Northport 5 & 10.

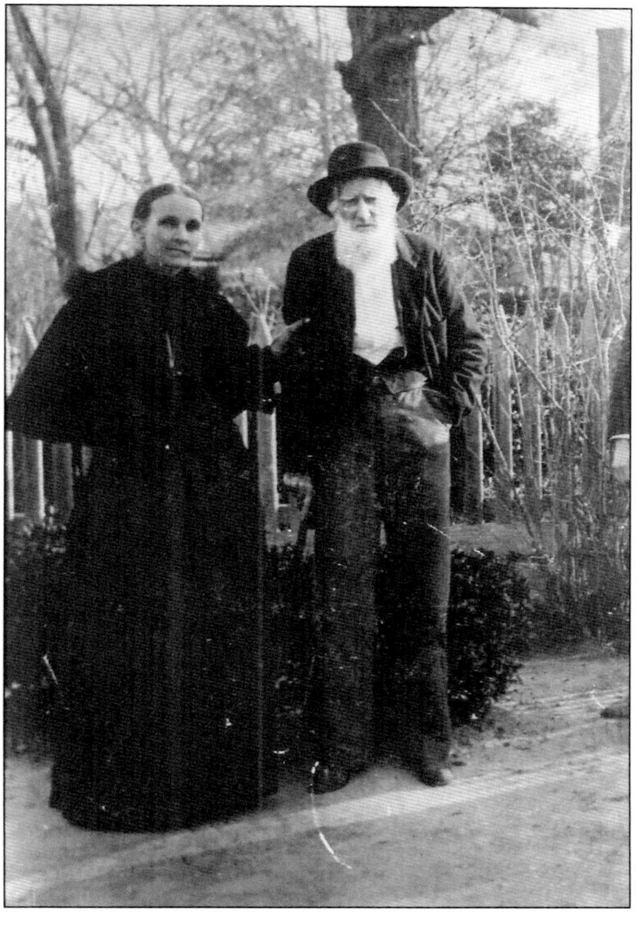

Seen here around 1900 are John Washington Deason and Frances Bell Deason. He was a leading merchant in Northport and a private in the 20th Alabama Company K in the Civil War. He was taken prisoner while on scouting duty near Chattanooga, Tennessee, in November 1863, and released after the conclusion of the war. In their later years, the Deasons lived in Dothan, Alabama.

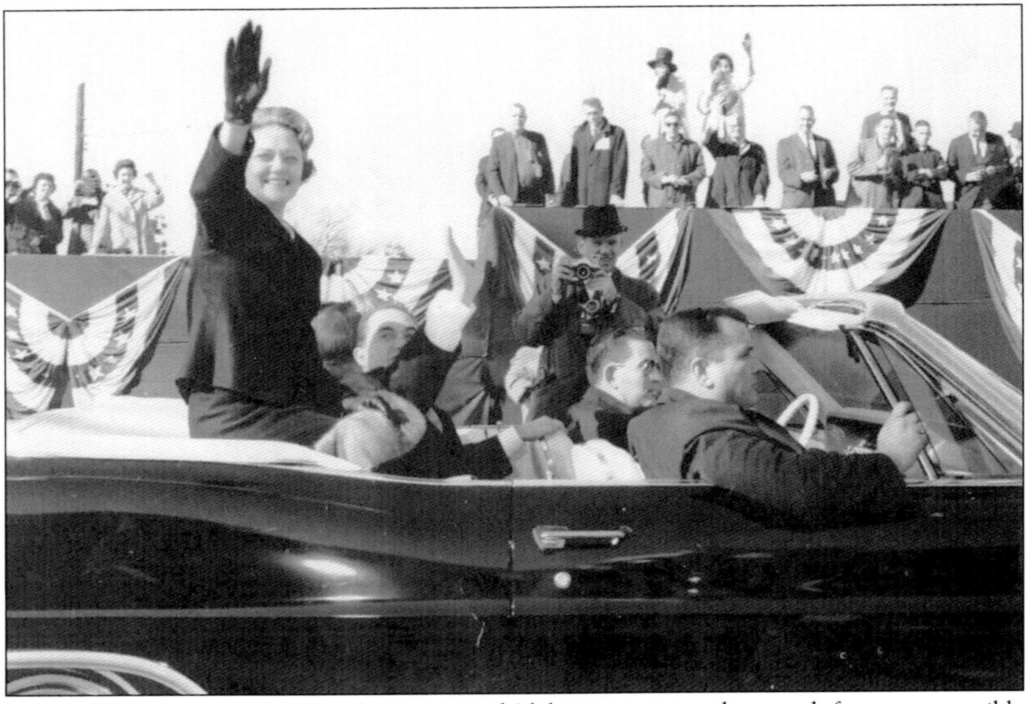

Lurleen B. Wallace, the first female governor of Alabama, waves to the crowds from a convertible during a 1966 parade. She was a 1942 graduate of Tuscaloosa County High School and a native of Northport. Wallace died May 7, 1968, after a long battle with cancer.

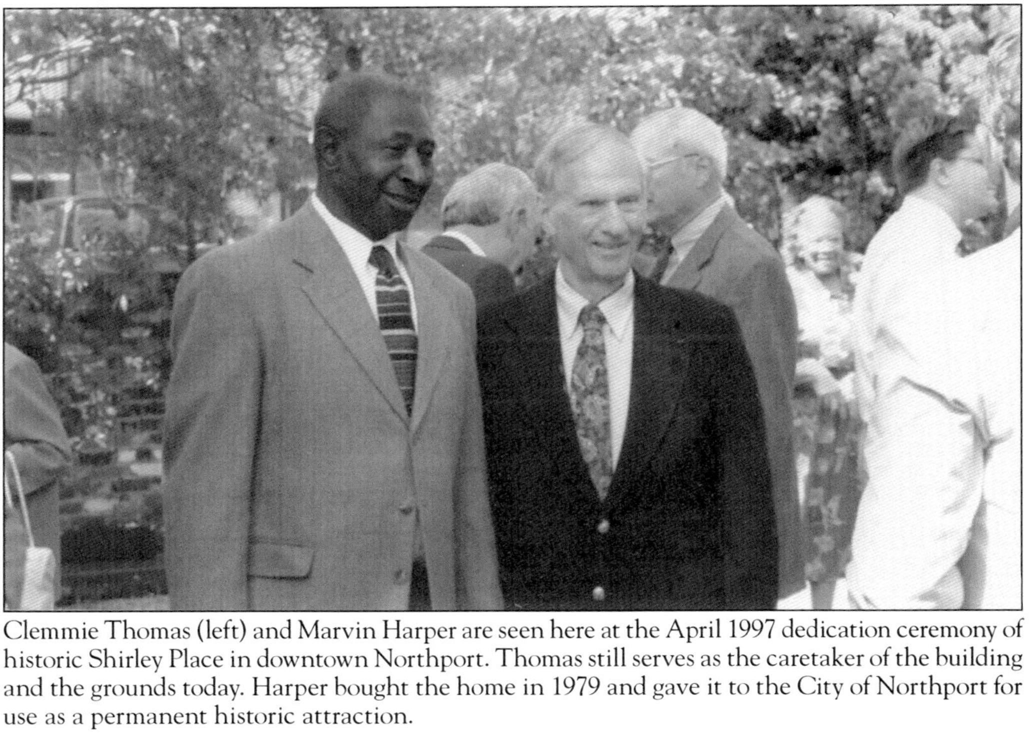

Clemmie Thomas (left) and Marvin Harper are seen here at the April 1997 dedication ceremony of historic Shirley Place in downtown Northport. Thomas still serves as the caretaker of the building and the grounds today. Harper bought the home in 1979 and gave it to the City of Northport for use as a permanent historic attraction.

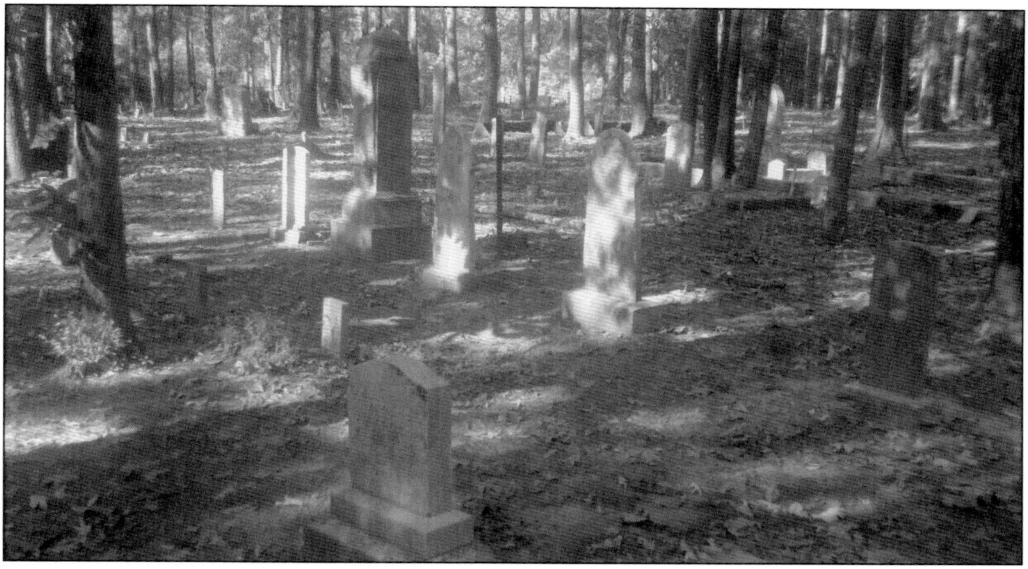

The historic Robertson-Stone Cemetery, also known as the Old Northport Cemetery, is the oldest cemetery in Northport. The cemetery was established in 1821 as the burial place for Catherine Murchison Findley. Among the types of grave markers are tent-tombs and brick-lined graves. After an extensive restoration effort headed by Lackey Stephens in 2008, the entire cemetery has been cleaned and mapped and old tombstones have been repaired. To date, the graves of more than 1,000 Northport citizens have been located.

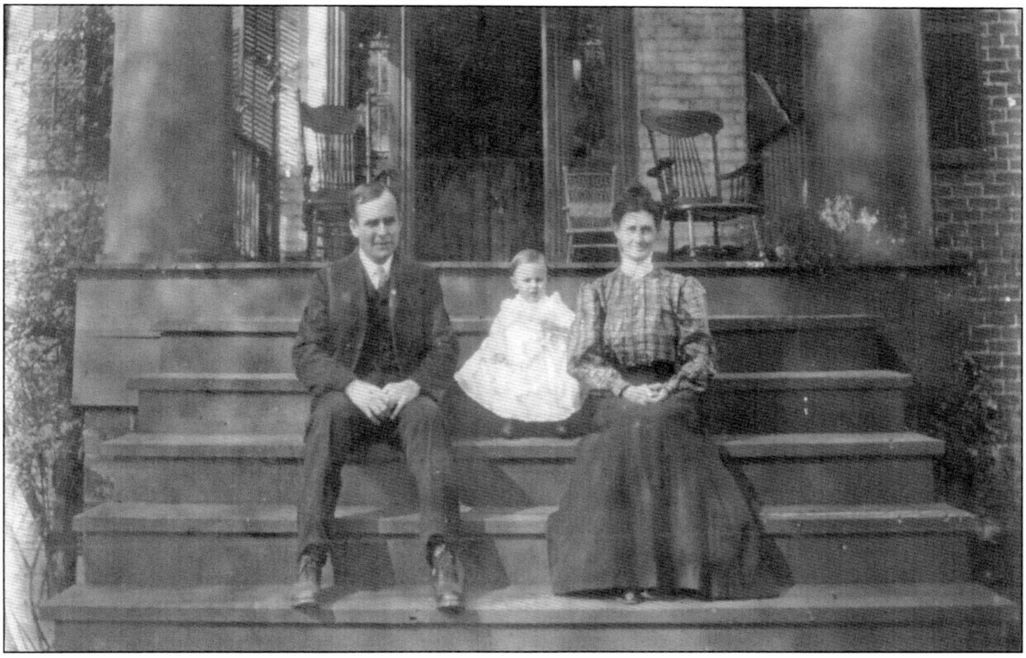

Sitting on the front steps of what was originally the G.W. Christian Home and is now historic Shirley Place are H. Clinton Deason, his wife, Lela, and their son, George W. Deason. George was born in September 1906, making the approximate date of this photograph spring 1907. Clinton Deason worked as a salesman in a general store in Northport.

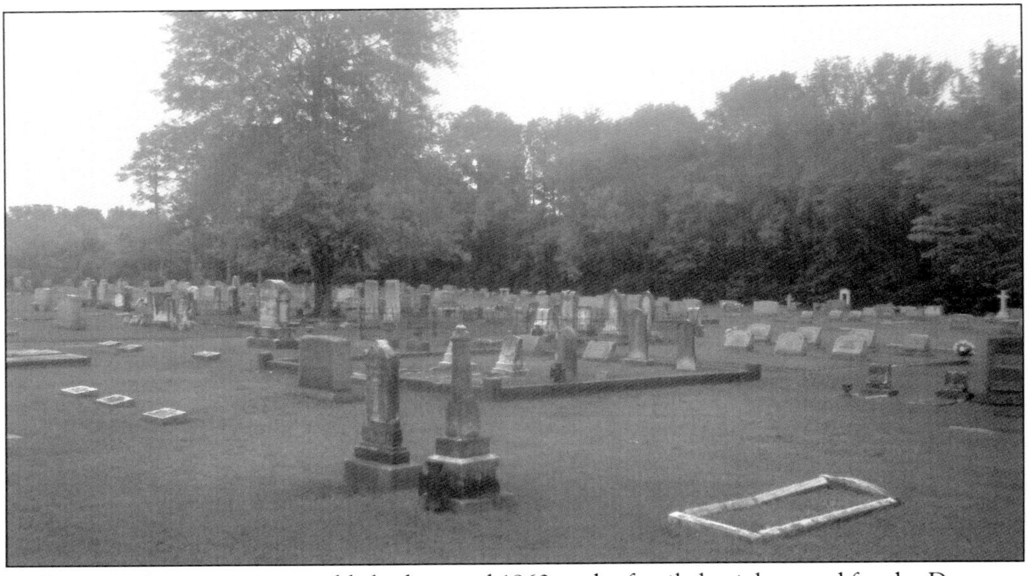

Williamson Cemetery was established around 1860 as the family burial ground for the Dempsey and Rebecca Williamson family. It is located on Fifth Street (formerly Columbus Street) a few blocks west of Main Avenue. Since 1887, the cemetery has been operated by the City of Northport. It was later developed in the urban garden style that became popular in the late 1800s.

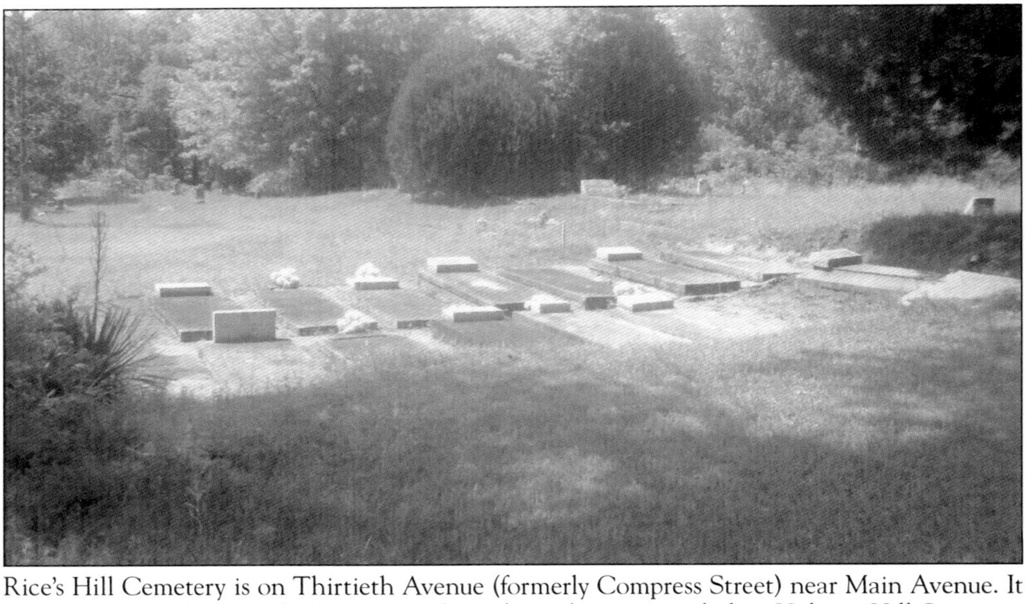

Rice's Hill Cemetery is on Thirtieth Avenue (formerly Compress Street) near Main Avenue. It has been known by several other names throughout the years, including Holman Hill Cemetery and Van Hoose Cemetery. Although the cemetery is near Mt. Galilee Baptist Church, it is not affiliated directly with the church.

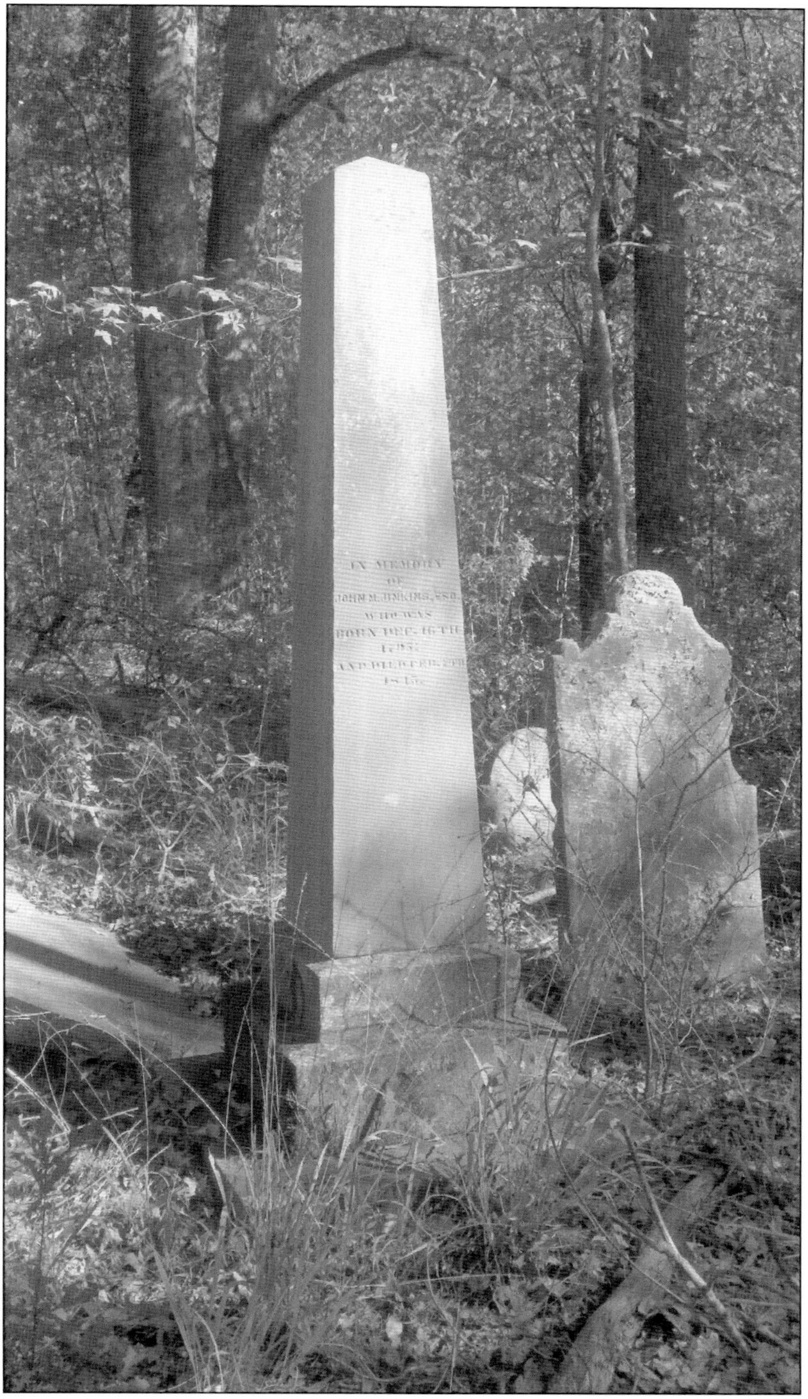

Jenkins Cemetery, on Highway 43 (formerly Byler Road) about a half-mile from Northport City Hall, has been in use since the 1830s. The Jenkins family settled in Tuscaloosa County before 1820. The cemetery has a special flat-slab tablet with information about the connection the Jenkins family had with the Methodist church for 30 years.

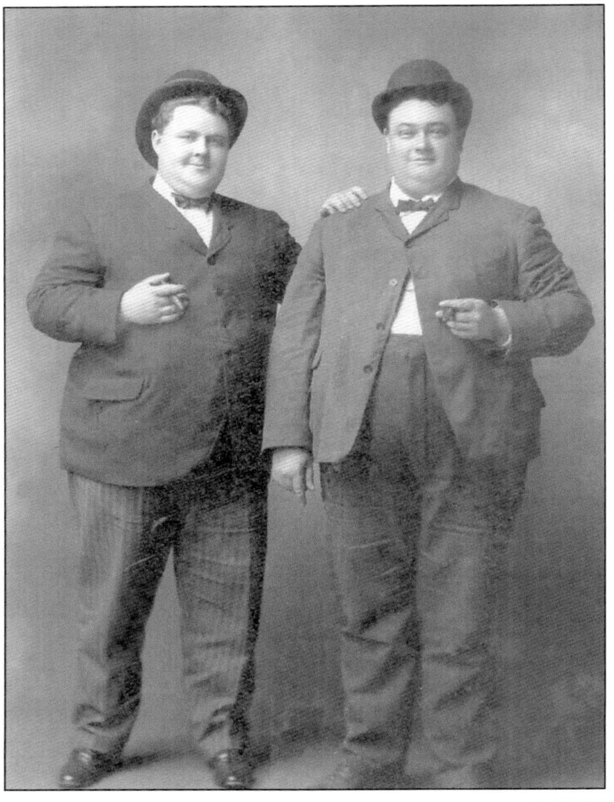

Charles Robert Maxwell (right) is seen here with a friend. His father, Richard Maxwell, came to this area in the 1840s from Cockermouth, England, with three brothers. Charles Maxwell was a successful merchant and a commercial fertilizer manufacturer. He and his wife, Julia Bell Maxwell, lived at the corner of Main Avenue and Park Street in an 1890s Victorian home. (Courtesy of Julia Maxwell Hallman, gifted to the Northport Heritage Museum.)

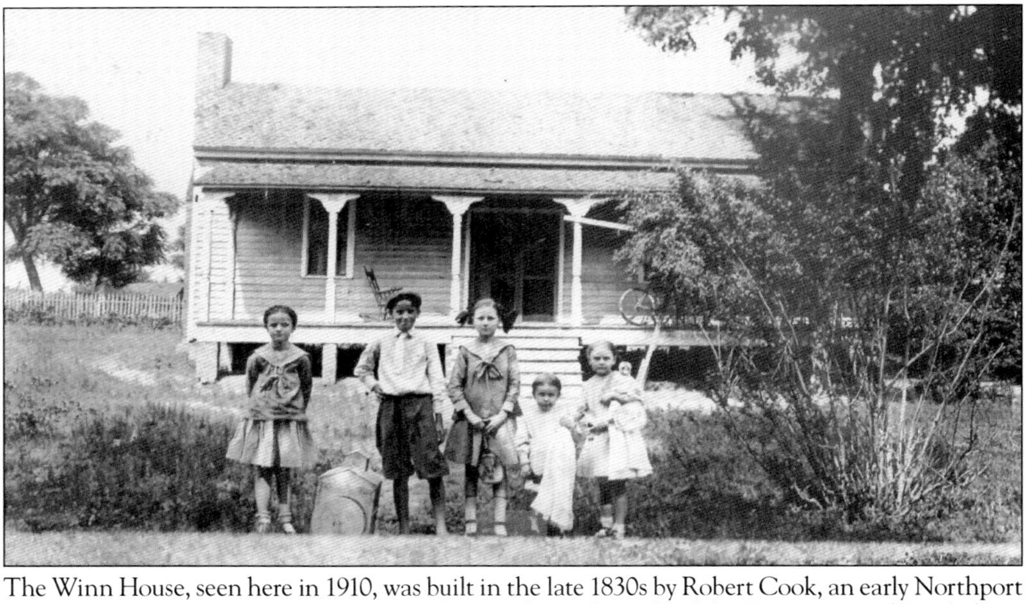

The Winn House, seen here in 1910, was built in the late 1830s by Robert Cook, an early Northport settler. It was located two lots north of the Northport branch of the First National Bank (now Doug Hollyhand Realty) on Main Avenue. Cook later built a home behind the Methodist church. Seen here are, from left to right, Genie Strong Freeman, Fred Strong, Mary Alice Strong Hinton, Wilma Strong Smith, and Reubel Strong Hamner.

Six

Bridging the Black Warrior

Beginning in 1834, seven bridges have crossed the Black Warrior River between Northport and Tuscaloosa. According to Ken Willis in his book *Spanning the Black Warrior River*, the first two bridges were town double-lattice covered bridges built by Seth King. The third bridge was built by Horace King, a freed slave who built many bridges in the southeast. The fourth bridge, a wrought iron bowstring design, was built by Zenas King of the King Iron & Bridge Company of Cleveland, Ohio.

It wasn't until the fifth bridge that someone without the surname King would build a bridge across the river. In 1895, James A. Whitner of Knoxville, Tennessee, was given the contract to build a four-span Parker truss or "camelback" truss bridge to replace the 1882 bridge. A new bridge was authorized by the US War Department that would meet new requirements set forth by the government.

By 1920, trucks crossing the river were exceeding the weight limits of the bridge. In 1922, a Waddell vertical-lift drawbridge was built by the Virginia Bridge and Iron Company. The new bridge could easily handle heavier commercial traffic.

In 1968, a six lane, 2,800-foot bridge made of concrete and steel I-beam girders was ordered to replace the aging drawbridge. The bridge was named for Hugh R. Thomas, a state representative from Tuscaloosa County who was killed in an automobile accident in 1967. The Hugh R. Thomas Bridge was dedicated in December 1973.

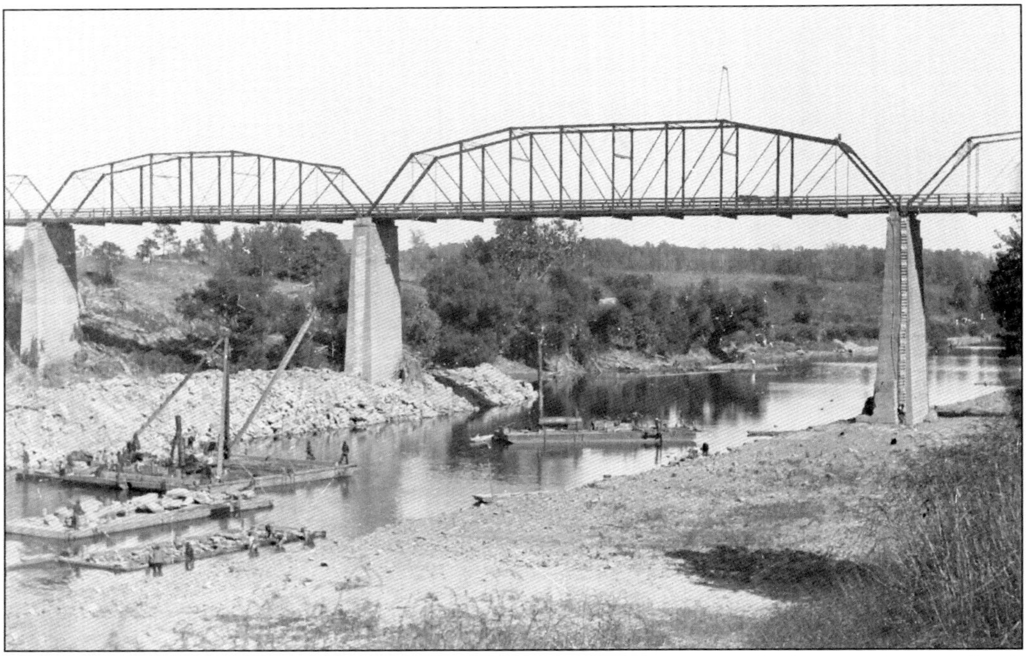

The 1897 Warrior River Bridge between Northport and Tuscaloosa was a Parker truss (or camelback truss) bridge, designed by Charles H. Parker. One span was built to swing open to allow passage for larger vessels and to relieve flooding during high water conditions. Below the bridge, workers dredge the shoals in order to improve the navigation channel.

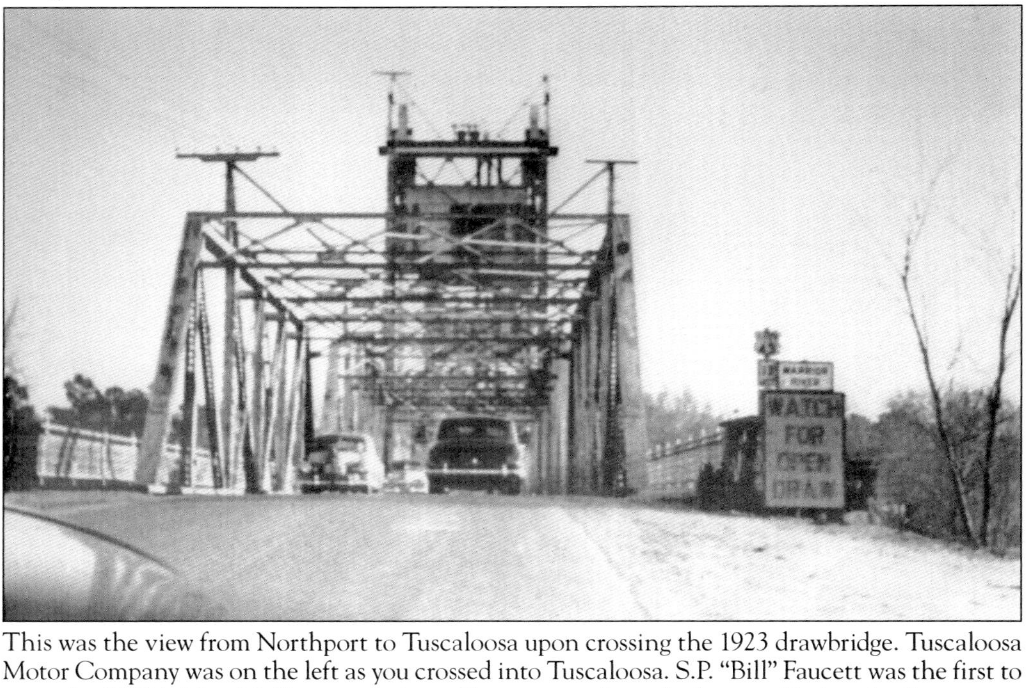

This was the view from Northport to Tuscaloosa upon crossing the 1923 drawbridge. Tuscaloosa Motor Company was on the left as you crossed into Tuscaloosa. S.P. "Bill" Faucett was the first to cross the 1923 bridge. Half a century later, Faucett was given the honor of being the first person to cross the new Hugh Thomas Bridge.

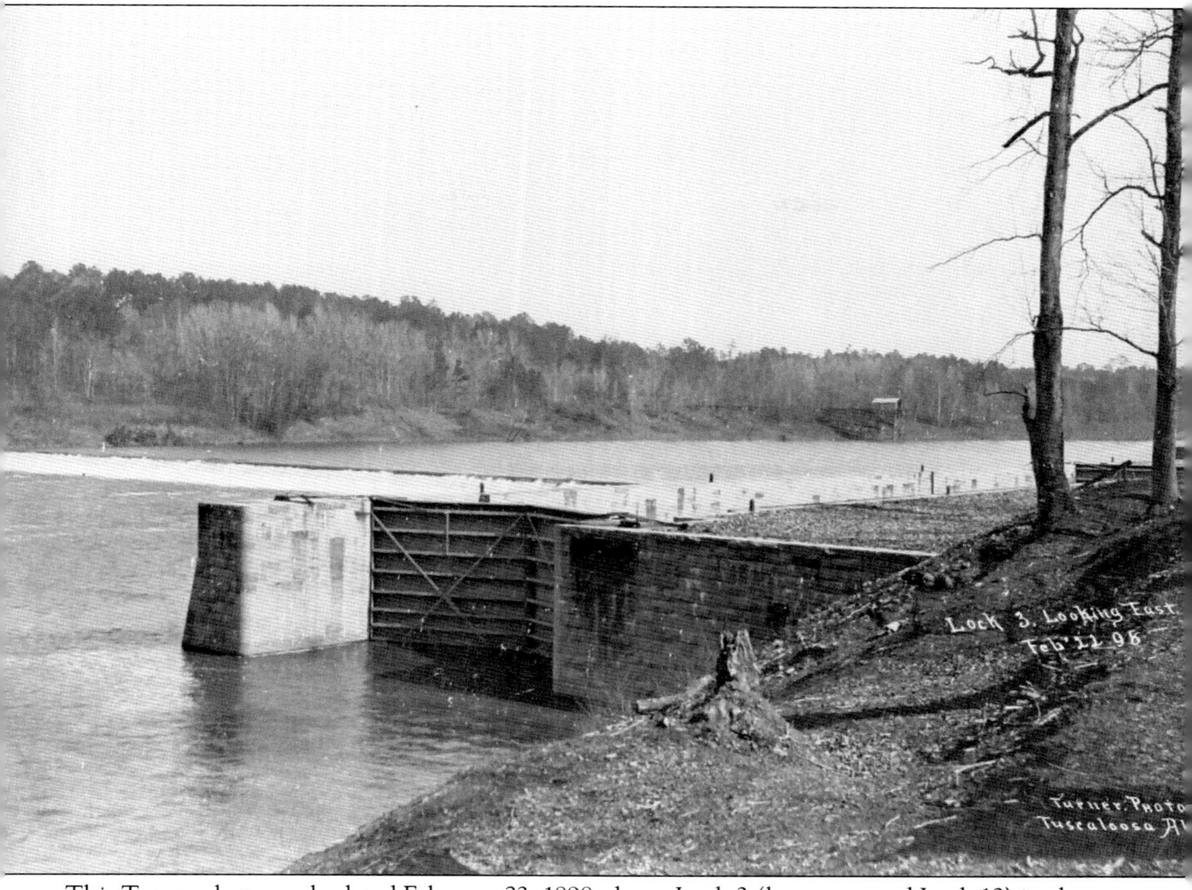

This Turner photograph, dated February 22, 1898, shows Lock 3 (later renamed Lock 12) in the foreground. Today, the remains of Lock 3 serve as part of The Park at Manderson Landing on the Jack Warner Parkway. In the background is the coal chute for Rice's mine. Cypress Inn Restaurant is now located at the opening of the old mine chute.

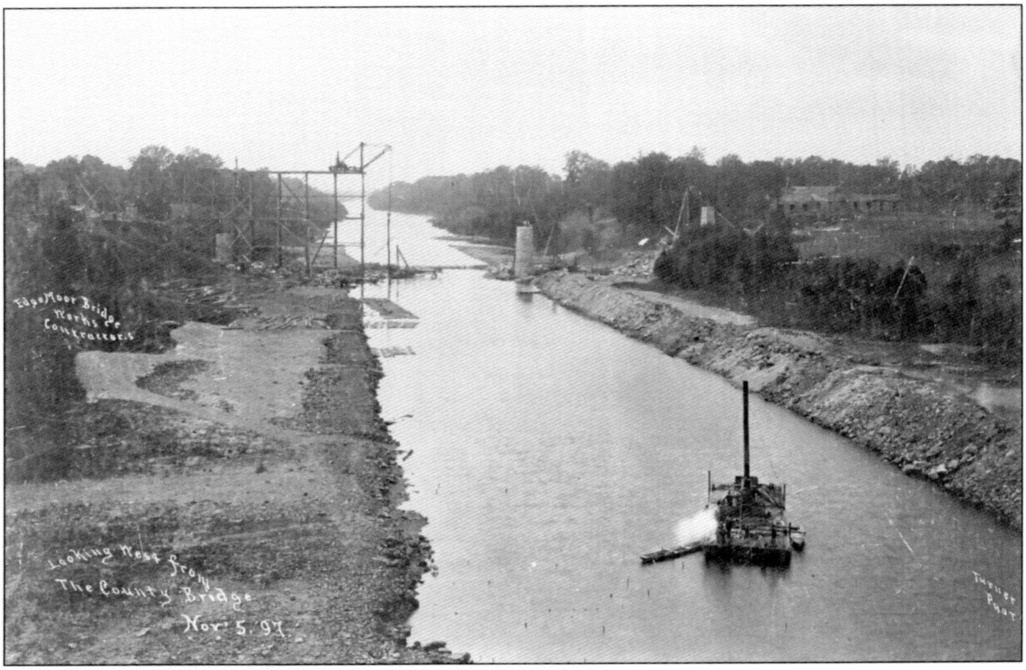

Looking west along the Black Warrior River, this Turner photograph was taken on November 5, 1897, from the 1896 river bridge. The Mobile & Ohio train bridge is under construction. To the right, along Northport's riverfront, is a cotton warehouse. The completion of the lock system along the Black Warrior River allowed the water level to rise for passage of larger vessels.

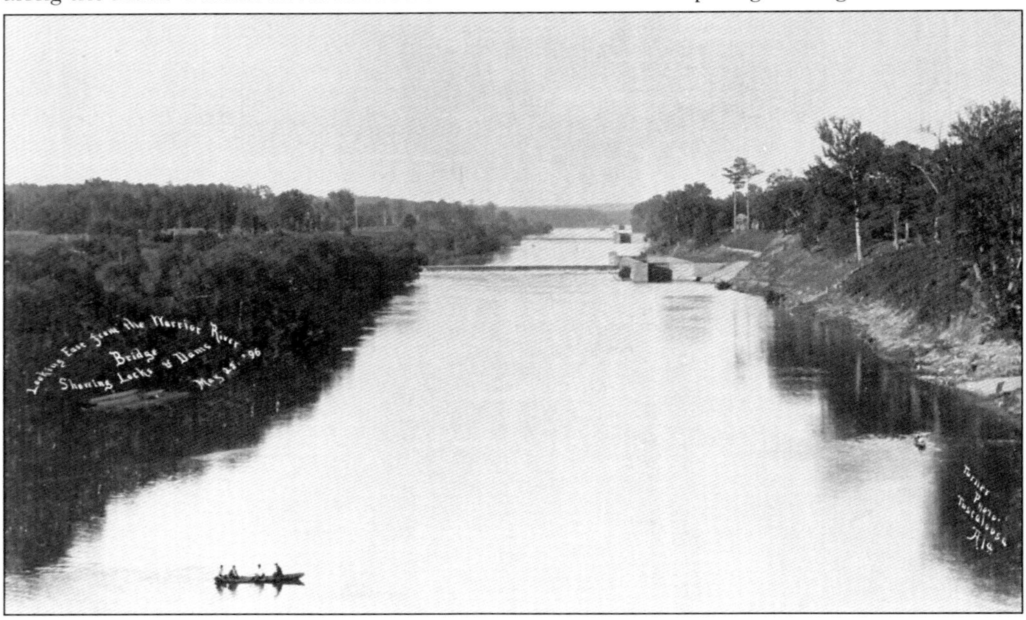

Taken on May 25, 1896, this Turner photograph shows Locks 1, 2, and 3 (later renamed Locks 10, 11, and 12) looking east from the river bridge. Part of Northport's riverfront is seen to the left. The completion of the lock and dam system along the Black Warrior River was mandated by the US War Department in 1875.

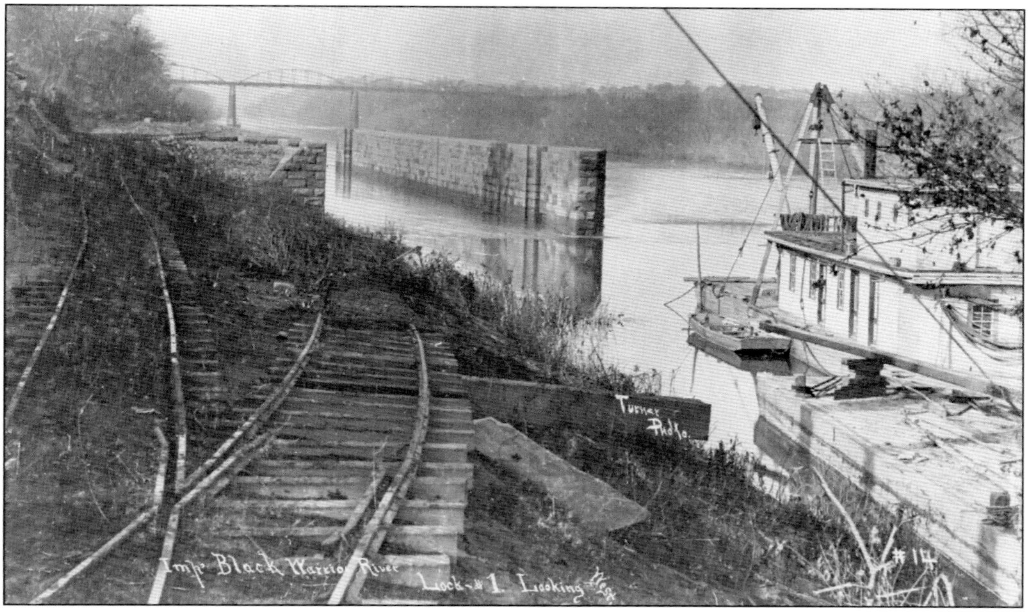

In the foreground of this Turner photograph is Lock 1 (later renamed Lock 10). In the background is the 1882 King Bowstring Bridge. Consisting of four wrought-iron spans, this was the first metal bridge to cross the Black Warrior River between Northport and Tuscaloosa. It was built by the King Iron & Bridge Company of Cleveland, Ohio.

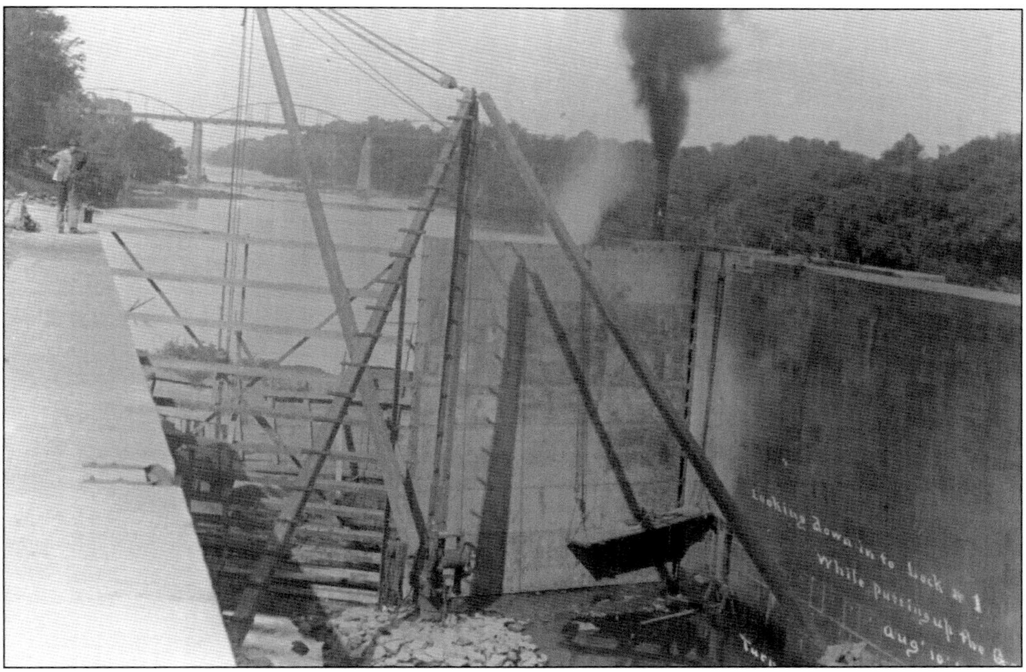

This 1890s Turner photograph shows the inside of Lock 1 while the gates were being put into place for permanent operation. In the background is the 1882 King Bowstring Bridge. The shoals of the river are seen just beyond the river crossing. Before the introduction of the river bridges, travelers could easily cross the river at the shoals, especially during the winter months.

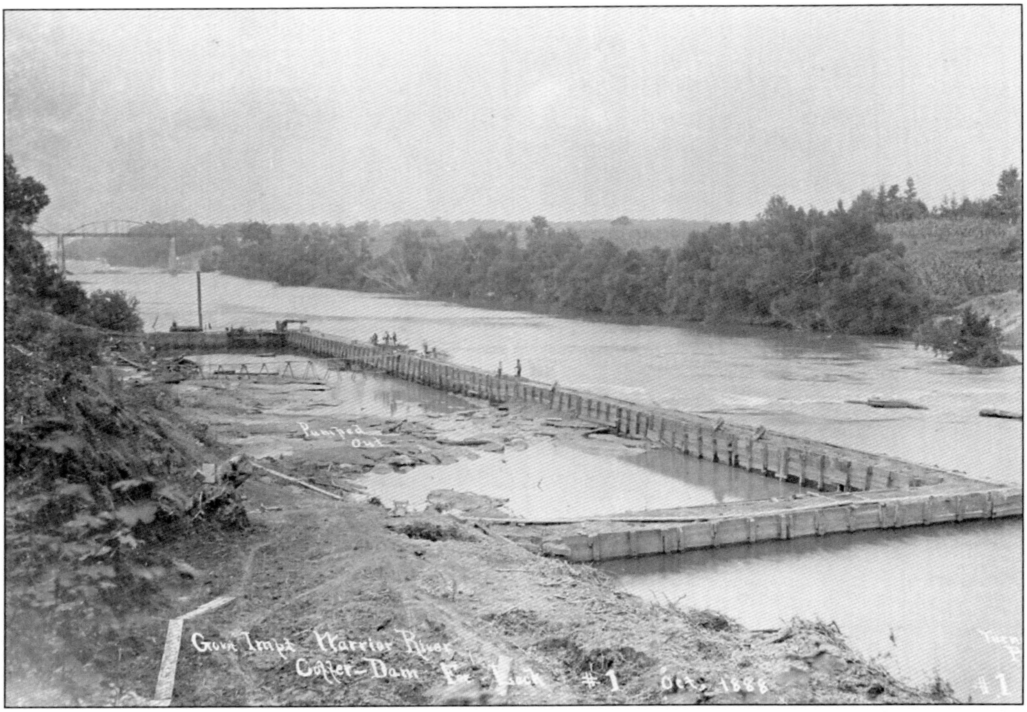

In the foreground of this October 1888 Turner photograph is the coffer dam for Lock 1. Once the coffer dam was complete, water was pumped out to allow for the lock walls to be put into place. In the background is the 1882 King Bowstring Bridge and Northport's riverfront.

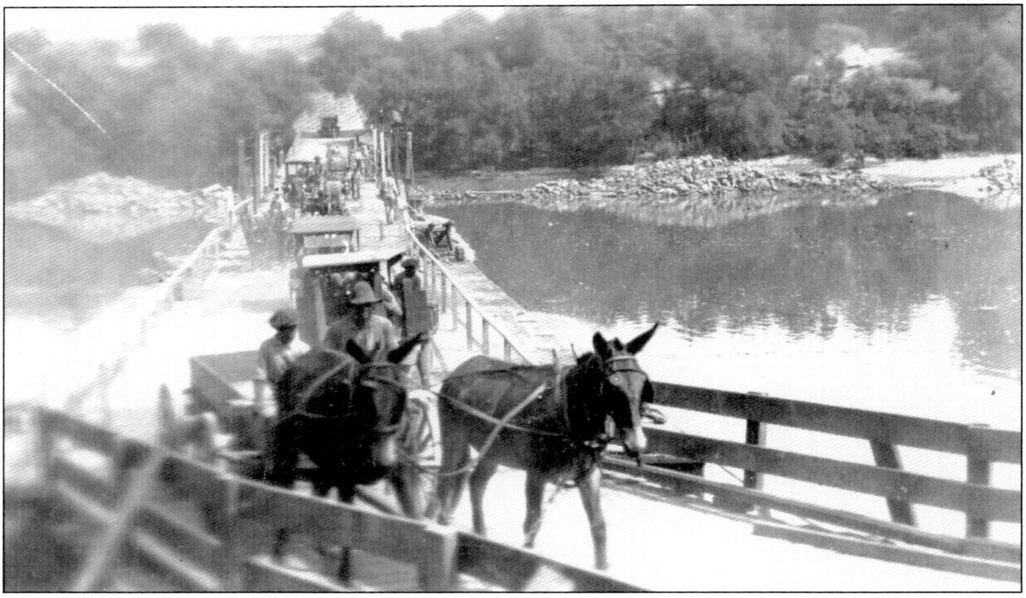

A pontoon bridge over the Black Warrior River was used from 1921 to 1922 to allow traffic to cross while the 1923 drawbridge was under construction. This photograph shows the northern view from the Tuscaloosa side of the river, where the Tuscaloosa Amphitheater is located today. The bridge led to a temporary road through the cotton gin to Main Avenue in Northport.

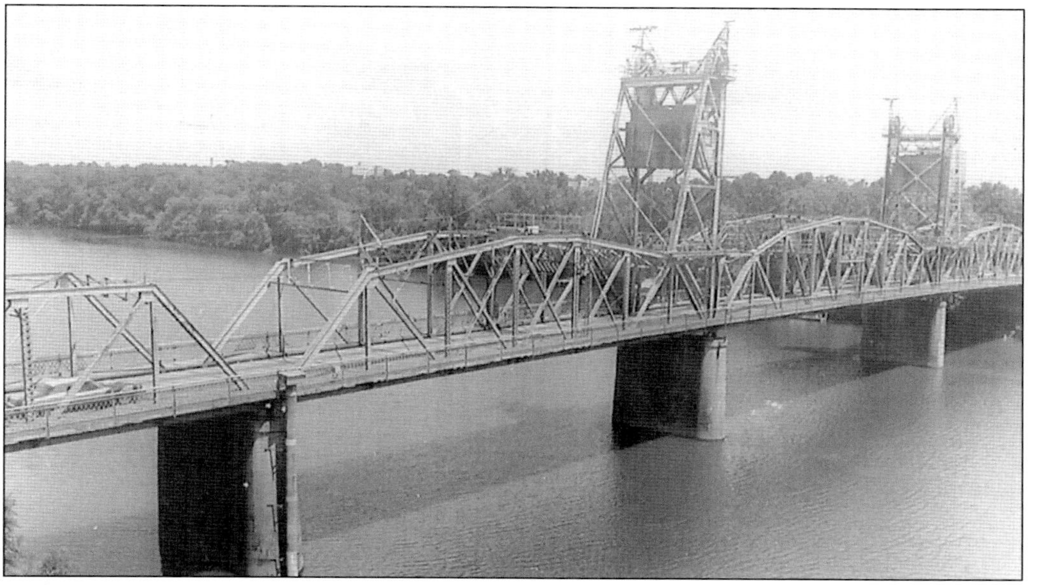

Built in 1923, the Waddell vertical lift drawbridge that crossed the Black Warrior River was destroyed in March 1974, after construction had begun on the new Hugh Thomas Bridge in December 1973. Many locals remember the old bridge, yet there are many in Northport today who have never heard of it. It was the last of its kind—the last to span the same spot where bridges had been built since 1835.

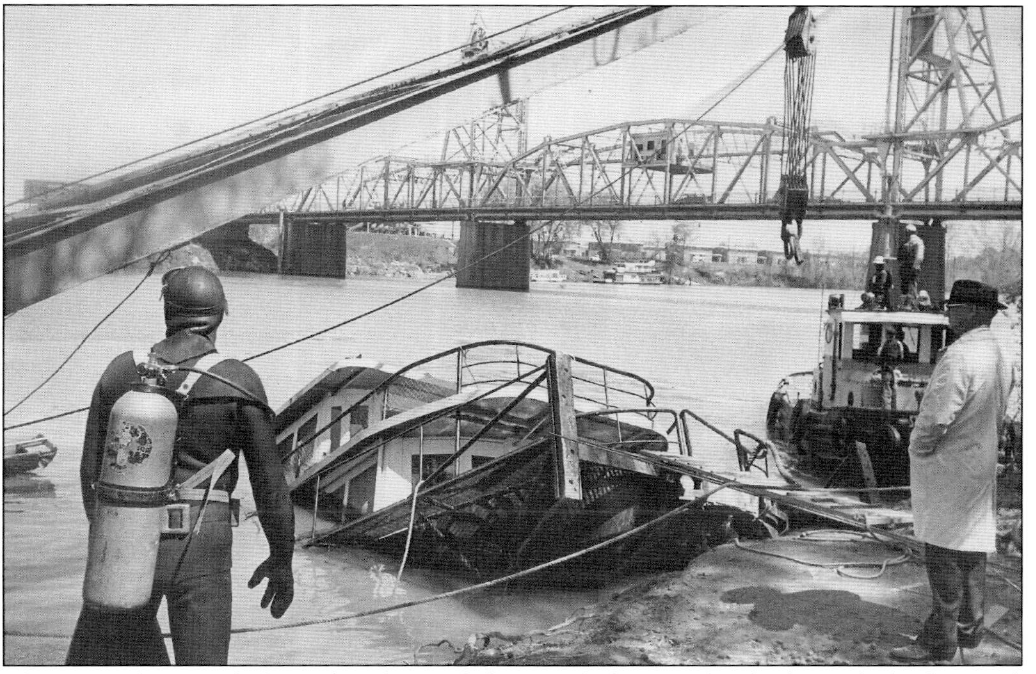

This 1960s photograph shows the salvage of a boat, with the 1923 drawbridge in the background. Before the Northport Fire Department, Northport businesses and residents depended on the Tuscaloosa Fire Department, as they did here. Northport had an agreement with Tuscaloosa to provide this service; in return, Northport provided support to the Tuscaloosa Fire Department.

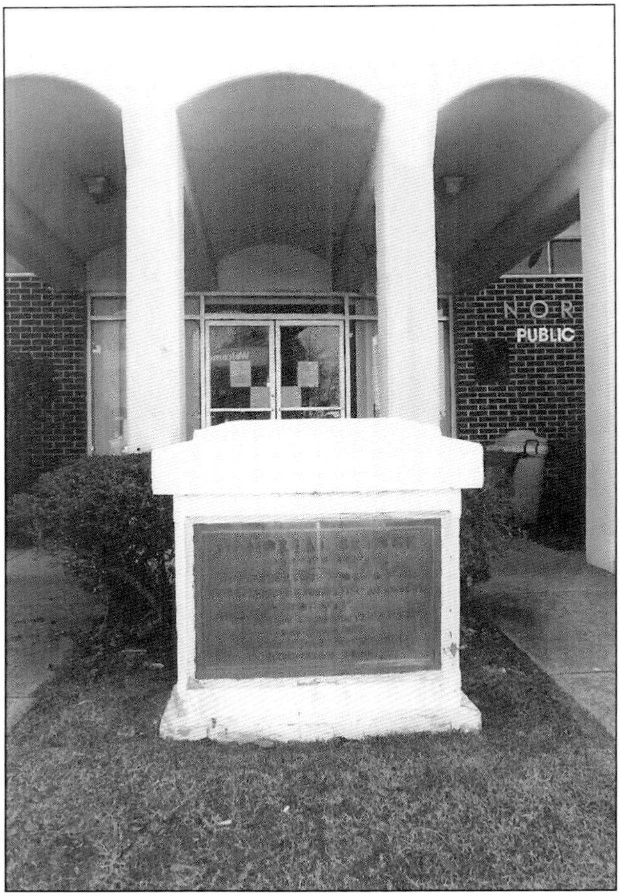

The Memorial Bridge marker was placed at the spot of the 1923 drawbridge and was dedicated in 1928. The marker reads "Memorial Bridge, Erected 1922, To the Memory of World War Soldiers from Tuscaloosa County, Affectionately Donated by the Northport Study Club, Dedicated 1928." In 1974, the marker was moved to Northport City Hall on Bridge Avenue. Today, the Memorial Bridge marker is located at the Van de Graaff Arboretum and Historic Bridge Park near downtown Northport.

This photograph was taken around 1898, shortly after the completion of the M&O train bridge and trestle. The crew is working on the Tuscaloosa side of the river. Downtown Northport, seen in the background, was saved when the train trestle was built in an s-curve. This design allowed its path to turn slightly west and away from the town center.

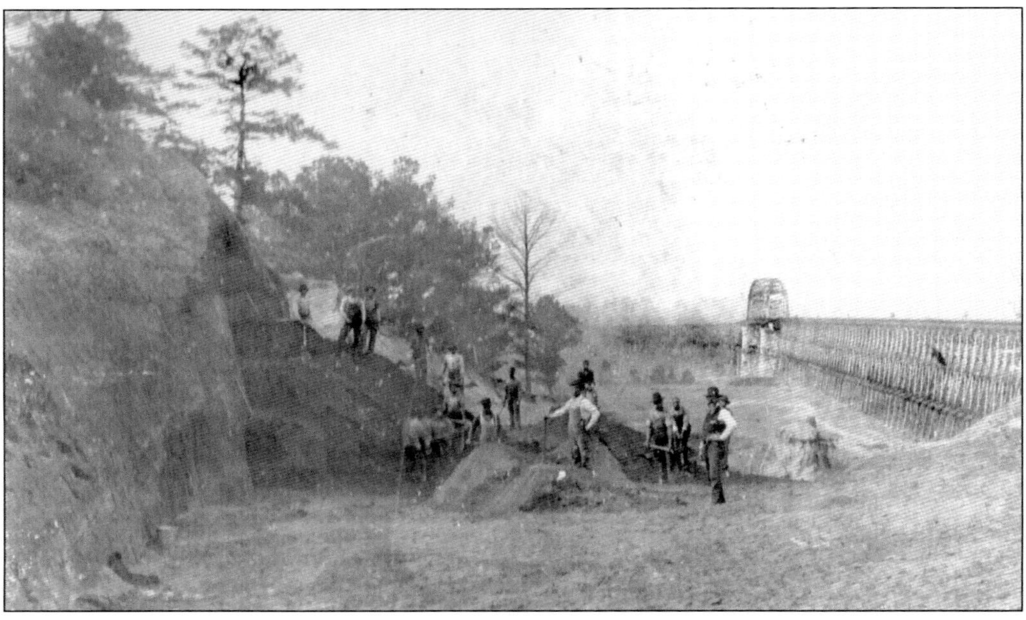

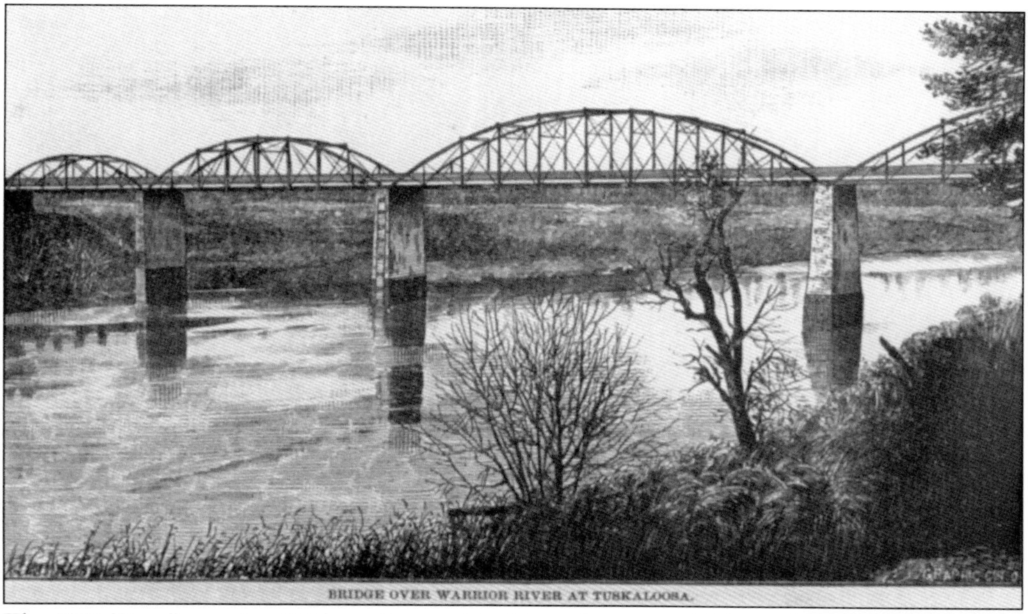

The 1882 King Bowstring Bridge is featured in this early postcard. The end of the bridge leading into Northport is on the left. Construction of the bridge began in December 1881 and the bridge was completed on Christmas Day, 1882. After 13 years of service, a new bridge was ordered built by the US government, which required that it be a draw-span bridge to accommodate increased river traffic. The 1882 bridge was dismantled starting in 1895. Each span was relocated to various locations around Tuscaloosa County for use as single-span bridges.

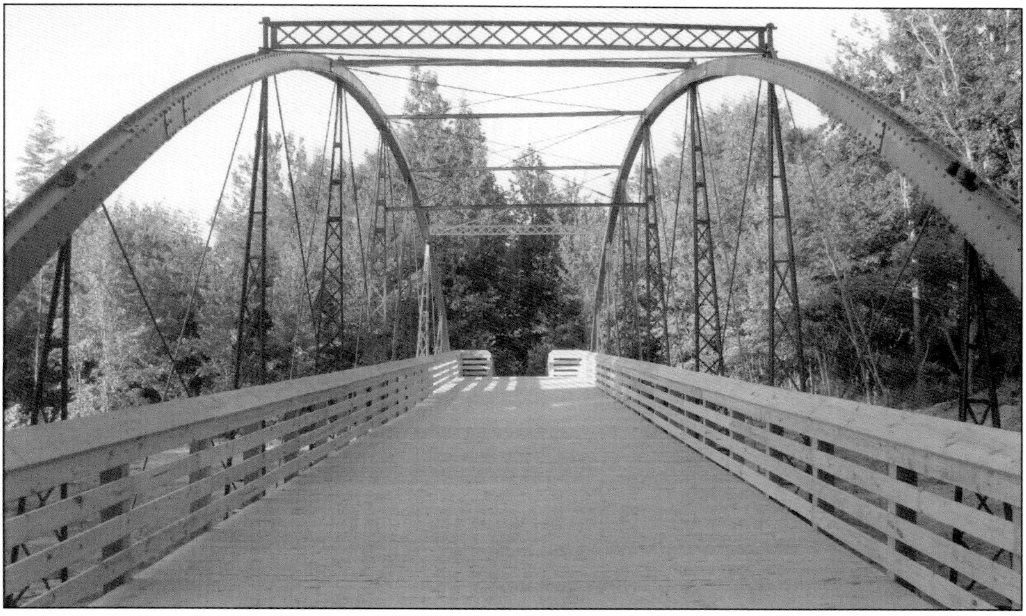

In August 2008, Friends of Historic Northport began the process of removing the only surviving span of the 1882 King Bowstring Bridge back to Northport as part of a bridge park. The generous donation of a 200-acre tract of land near downtown Northport by the Jemison–Van de Graaff–Rountree family allowed for the completion of the 1882 Bowstring Bridge Project in September 2010.

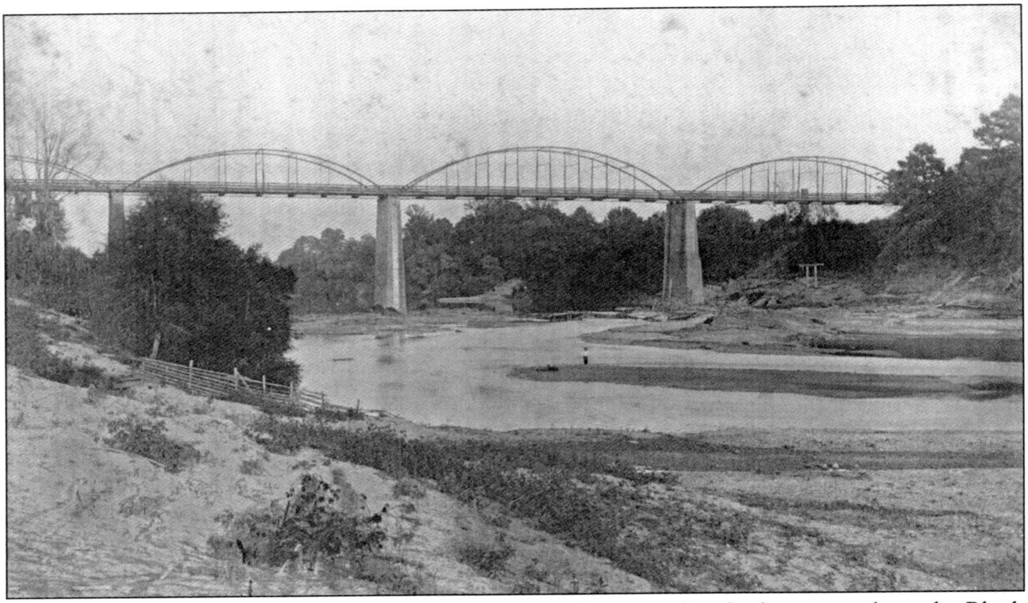

This early photograph of the 1882 King Bowstring Bridge was taken looking east along the Black Warrior River from the Marr's Ferry Road landing. This historic road, built by William Marr around 1830, still exists as it did in the 1830s from the banks of the river to the levee property just west of the train trestle. William Marr operated a ferry between Northport and New Town. (Courtesy of the W.S. Hoole Special Collections Library, University of Alabama.)

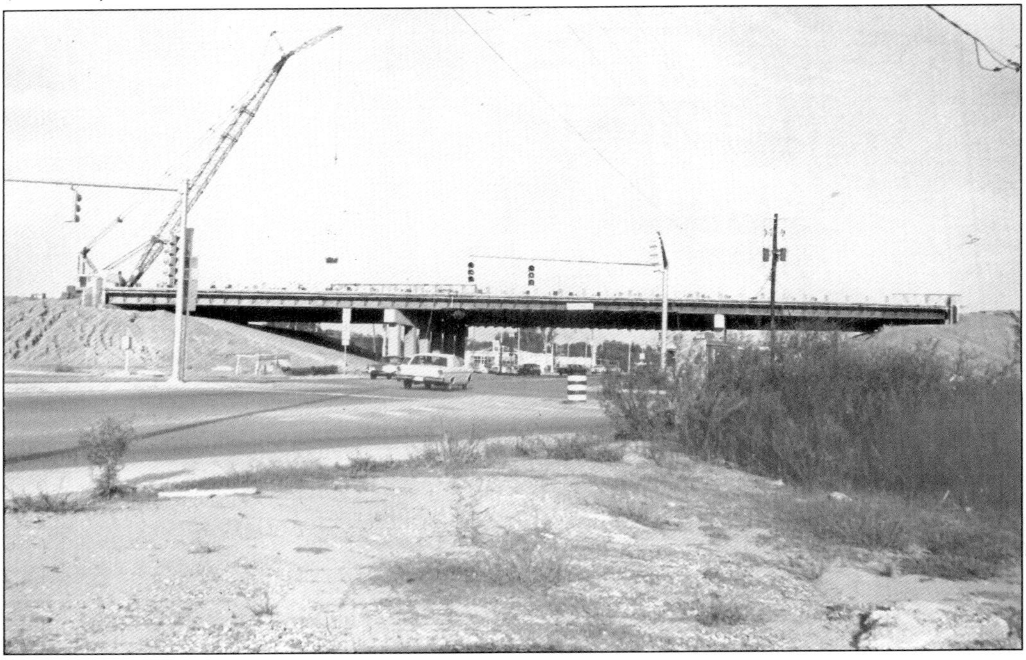

This c. 1970 photograph was taken looking east from downtown Northport toward the construction of the Fifth Street overpass and ramps connecting Lurleen B. Wallace Boulevard to the new Hugh Thomas Bridge. More than 40 years later, thousands of vehicles use this overpass every day to cross to and from Tuscaloosa.

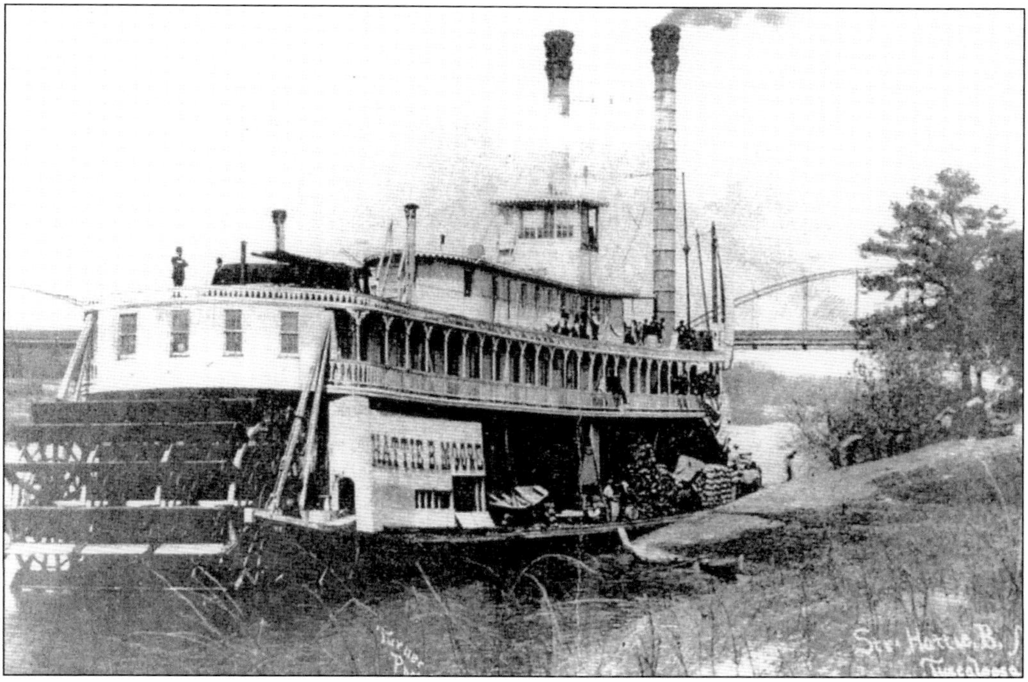

The *Hattie B. Moore* was a riverboat that navigated along the Black Warrior River in the late 1800s and early 1900s. The 1882 King Bowstring Bridge is in the background. (Courtesy of Heritage Commission of Tuscaloosa County.)

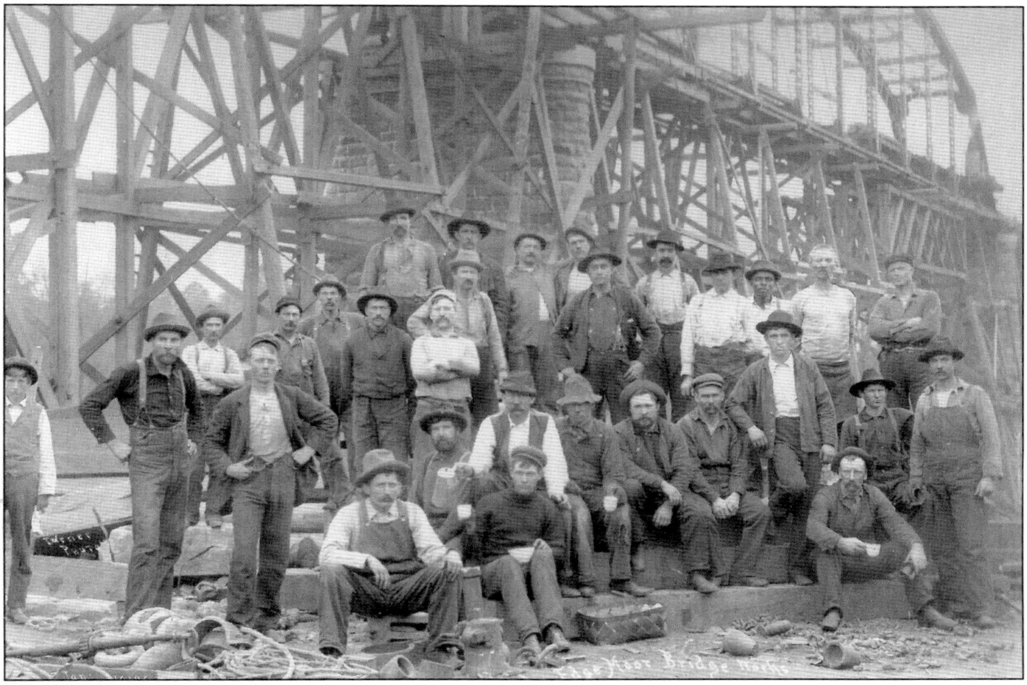

This January 10, 1898, photograph shows workers and builders employed by Edgemoor Bridge Works stopping for a photograph during the construction of the M&O river bridge.

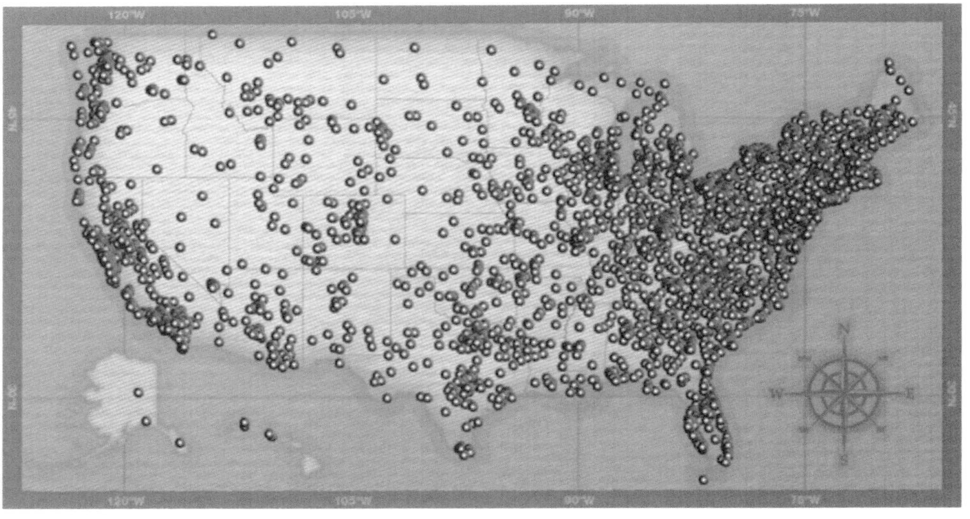